TABLE OF CONTENTS

PERIODS OF CREATION AND RECORDS OF CREATION

The revolutionary influence which the Theory of Descent must exercise upon all sciences, will in all probability affect no branch of science, excepting Anthropology, so much as the descriptive portion of natural history, that which is known as systematic Zoology and Botany. Most naturalists who have hitherto occupied themselves with arranging the different systems of animals and plants, have collected, named, and arranged the different species of these natural bodies with much the same interest as antiquarians and ethnographers collect the weapons and utensils of different nations. Many have not even risen above the degree of intelligence with which people usually collect, label, and arrange crests, stamps, and similar curiosities. In the same manner as some collectors find their pleasure in the similarity of forms, the beauty or rarity of the crests or stamps, and admire in them the inventive art of man, so many naturalists take a delight in the manifold forms of animals and plants, and marvel at the rich imagination of the Creator, at His unwearied creative activity, and at His curious fancy for forming, by the side of so many beautiful and useful organisms, also a number of ugly and useless ones.

This childlike treatment of systematic Zoology and Botany is completely annihilated by the Theory of Descent. In the place of the superficial and playful interest with which most naturalists have hitherto regarded organic structures, we now have the much higher interest of the intelligent understanding which detects in the related forms of organisms their true blood relationships. The Natural System of animals and plants, which was formerly valued either only as a registry of names, to facilitate the survey of

5

the different forms, or as a table of contents for the short expression of their degrees of similarity, receives from the Theory of Descent the incomparably higher value of a true pedigree of organisms. This pedigree is to disclose to us the genealogical connection of the smaller and larger groups. It has to show us in what way the different classes, orders, families, genera, and species of the animal and vegetable kingdoms correspond with the different branches, twigs, and groups of twigs of the pedigree. Every wider and higher category or stage of the system (for example a class, or an order) comprises a number of larger and stronger branches of the pedigree; every narrower and lower category (for example a genus, or a species) only a smaller and thinner group of twigs. It is only when we thus view the natural system as a pedigree that we perceive its true value. (Gen. Morph. ii. Plate XVII. p. 397.)

Since we hold fast this genealogical conception of the Organic System, to which alone undoubtedly the future of classificatory Zoology and Botany belongs, we should now turn our attention to one of the most essential, but also one of the most difficult, tasks of the "non-miraculous history of creation," namely, to the actual construction of the Organic Pedigree. Let us see how far we are already able to point out all the different organic forms as the divergent descendants of a single or of some few common original forms. But how can we construct the actual pedigree of the animal and vegetable group of forms from our knowledge of them, at present so scanty and fragmentary? The answer to this question lies in what we have already remarked of the parallelism of the three series of development—in the important causal relation which connects the palæontological development of all organic tribes with the embryological development of individuals, and with the systematic development of groups.

In order to accomplish our task we shall first have to direct our attention to palæontology, or the science of petrifactions. For if the Theory of Descent is really true, if the petrified remains of formerly living animals and plants really proceed from the extinct primæval ancestors and progenitors of the present organisms, then, without anything else, the knowledge and comparison of petrifactions ought to disclose to us the pedigree of organisms. However simple and clear this may seem in theory, the task becomes extremely hard and complicated when it is actually taken in hand. Its practical solution would be very difficult even if the petrifactions were to any extent completely preserved. But this is by no means the case. The obvious records of creation which lie buried in petrifactions are imperfect beyond all measure. Hence it is necessary critically to examine these records, and to determine the value which petrifactions possess for the history of the development of organic tribes. As I have previously discussed the general importance of petrifactions as the records of creation, when we were considering Cuvier's merits in the science of fossils, we may now at once examine the conditions and circumstances under which the remains of organic bodies became petrified and preserved in a more or less recognizable form.

As a rule we find petrifactions or fossils enclosed only in those stones which have been deposited in layers as mud by water, and which are on that account called neptunic, stratified, or sedimentary rocks. The deposition of such strata could of course only commence after the condensation of watery vapour into liquid water had taken place in the course of the earth's history. After that period, which we considered in our last chapter, not only did life begin on the earth, but also an uninterrupted and exceedingly important transformation of the rigid inorganic crust of the earth. The water began that extremely important mechanical action by which the surface of the earth is perpetually, though slowly, transformed. I may surely presume that it is generally known what an extremely important influence, in this respect, is even yet exercised by water at every moment. As it falls down as rain, trickling through the upper strata of the earth's crust, and flowing down from heights into hollows, it chemically dissolves different mineral parts of the ground, and mechanically washes away the loose particles. In flowing down from mountains water carries their debris into the plains, or deposits it as mud in stagnant lakes. Thus it continually works at lowering mountains and filling up valleys. In like manner the breakers of the sea work uninterruptedly at the destruction of the coasts and at filling up the bottom of the sea with the debris they wash down. The action of water alone, if it were not counteracted by other circumstances, would in time level the whole earth. There can be no doubt that the mountain masses—which are annually carried down as mud into the sea, and deposited on its floor—are so great that in the course of a longer or shorter period, say a few millions of years, the surface of the earth would be completely levelled and become enclosed by a continuous sheet of water. That this does not happen is owing to the perpetual volcanic action of the fiery-fluid centre of the earth. The surging of the melted nucleus against the firm crust necessitates continual alternations of elevation and depression on the different parts of the earth's surface. These elevations and depressions for the most part take place very slowly; but, as they continue for thousands of years, by the combined effect of small, interrupted movements, they produce results no less grand than does the counteracting and levelling action of water.

Since the elevations and depressions of the different parts of the earth alternate with one another in the course of millions of years, first this and then that part of the earth's surface is above or below the level of the sea. I have already given examples of this in the preceding chapter (vol. i. p. 361). Hence, in all probability, there is no part of the outer crust of the earth which has not been repeatedly above and also below the level of the sea.

This repeated change explains the variety and the different composition of the numerous neptunic strata of rocks, which in most places have been deposited one above another in considerable thickness. In the different periods of the earth's history during which these deposits took place there lived various and different populations of animals and plants. When their dead bodies sank to the bottom of the waters, the forms of the bodies impressed themselves upon the soft mud, and imperishable parts, such as hard bones, teeth, shells, etc., became enclosed in it uninjured. These were preserved in the mud, which condensed them into neptunic rock, and as petrifactions they now serve to characterise the respective strata. By a careful comparison of the different strata lying one above another, and the petrifactions preserved in them, it has become possible to decide the relative age of the strata and groups of strata, and to establish, by direct observation, the principal eras of phylogeny, that is to say, the stages in history of the development of animal and vegetable tribes.

The different strata of neptunic rocks deposited one above another, which are composed in very various ways of limestone, clay, and sand, geologists have grouped together into an ideal System or Series, which corresponds with the whole course of the organic history of the earth, or with that portion of the earth's history during which organic life existed. Just as so-called "universal history" falls into larger and smaller periods, which are characterized by the conditions of development of the most important nations at the respective epochs, and are separated from one another by great events, so we also divide the infinitely longer organic history of the earth into a series of greater and less periods. Each of these periods is distinguished by a characteristic flora and fauna, and by the specially strong development of certain vegetable or animal groups, and each is separated from the preceding and succeeding period by a striking change in the character of its animal and vegetable inhabitants.

In relation to the following survey of the historical course of development which the large animal and vegetable tribes have passed through, it will be desirable to say a few words first as to the systematic classification of the neptunic groups of strata, and the larger and smaller periods corresponding to them. As will be seen directly, we are able to divide the whole of the sedimentary rocks lying one above another into five main groups or periods, each period into several subordinate groups of strata or systems, and each system of strata again into still smaller groups or formations; finally, each formation can again be divided into stages or sub-formations, and each of these again into still smaller layers or beds. Each of the five great rock-groups was deposited during a great division of the earth's history, during a long era or epoch; each system during a shorter period; each formation during a still shorter period. In thus reducing the periods of the organic history of the earth, and the neptunic strata containing petrifactions deposited during those periods into a connected system, we proceed exactly like the historian who divides the history of nations into the three main divisions of Antiquity, the Middle Ages, and Modern Times, and each of those sections again into subordinate periods and epochs. But the historian by this sharp systematic division, and by fixing the boundary of the periods by particular dates, only seeks to facilitate his survey, and in no way means to deny the uninterrupted connection of events and the development of nations. Exactly the same qualification applies to our systematic division, specification, or classification of the organic history of the earth. Here, too, a continuous thread runs through the series of events unbroken. We must therefore distinctly protest against the idea that by sharply bounding the larger and smaller groups of strata, and the the the periods corresponding with them, we in any way wish to adopt Cuvier's doctrine of terrestrial revolutions, and of repeated new creations of organic populations. That this erroneous doctrine has long since been completely refuted by Lyell, I have already mentioned. (Compare vol. i. p. 127.)

The five great main divisions of the organic history of the earth, or the palæontological history of development, we call the primordial, primary, secondary, tertiary, and quaternary epochs. Each is distinctly characterized by the predominating development of certain animal and vegetable groups in it, and we might accordingly symbolically designate the five epochs, on the one hand by the names of the groups of the vegetable kingdom, and on the other hand by those of the different classes of vertebrate animals. In this case the first, or primordial epoch, would be the era of the Tangles (Algæ) and skull-less Vertebrates; the second, or primary epoch, that of the Ferns and Fishes; the third, or secondary epoch, that of Pine Forests and Reptiles; the fourth, or tertiary epoch, that of Foliaceous Forests and of Mammals; finally, the fifth, or quaternary epoch, the era of Man, and his Civilization. The divisions or periods which we distinguish in each of the five long eras (p. 14) are determined by the different systems of strata into which each of the five great rock-groups is divided (p. 15). We shall now take a cursory glance at the series of these systems, and at the same time at the populations of the five great epochs.

The first and longest division of the organic history of the earth is formed by the primordial epoch, or the era of the Tangle Forests. It comprises the immense period from the first spontaneous generation, from the origin of the first terrestrial organism, to the end of the Silurian system of deposits. During this immeasurable space of time, which in all probability was much longer than all the other four epochs taken together, the three most extensive of all the neptunic systems of strata were deposited, namely, the Laurentian, upon that the Cambrian, and upon that the Silurian system. The approximate thickness or size of these three systems together amounts to 70,000 feet. Of these about 30,000 belong to the Laurentian, 18,000 to the

Cambrian, and 22,000 to the Silurian system. The average thickness of all the four other rock groups, the primary, secondary, tertiary, and quaternary, taken together, may amount at most to 60,000 feet; and from this fact alone, apart from many other reasons, it is evident that the duration of the primordial period was probably much longer than the duration of all the subsequent periods down to the present day. Many thousands of millions of years were required to deposit such masses of strata. Unfortunately, by far the largest portion of the primordial group of strata is in the metamorphic state (which we shall directly explain), and consequently the petrifactions contained in them—the most ancient and most important of all—have, to a great extent, been destroyed and become unrecognisable. Only in one portion of the Cambrian strata have petrifactions been preserved in a recognizable condition and in large quantities. The most ancient of all distinctly preserved petrifactions has been found in the lowest Laurentian strata (in the Ottawa formation), which I shall afterwards have to speak of as the "Canadian Life's-dawn" (Eozoon canadense).

Although only by far the smaller portion of the primordial or archilithic petrifactions are preserved to us in a recognizable condition, still they possess the value of inestimable documents of the most ancient and obscure times of the organic history of the earth. What seems to be shown by them, in the first place, is that during the whole of this immense period there existed only inhabitants of the waters. As yet, at any rate, among all archilithic petrifactions, not a single one has been found which can with certainty be regarded as an organism which has lived on land. All the vegetable remains we possess of the primordial period belong to the lowest of all groups of plants, to the class of Tangles or Algæ, living in water. In the warm primæval sea, these constituted the forests of the period, of the richness of which in forms and density we may form an approximate idea from their present descendants, the tangle forests of the Atlantic Sargasso sea. The colossal tangle forests of the archilithic period supplied the place of the forest vegetation of the mainland, which was then utterly wanting. All the animals, also, whose remains have been found in archilithic strata, like the plants, lived in water. Only crustacea are met with among the animals with articulated feet, as yet no spiders and no insects. Of vertebrate animals, only a very few remains of fishes are known as having been found in the most recent of all primordial strata, in the upper Silurian. But the headless vertebrate animals, which we call skull-less, or Acrania, and out of which fishes must have been developed, we suppose to have lived in great numbers during the primordial epoch. Hence we may call it after the Acrania as well as after the Tangles.

The primary epoch, or the era of Fern Forests, the second main division of the organic history of the earth, which is also called the palæolithic or palæozoic period, lasted from the end of the Silurian formation of strata to the end of the Permian formation. This epoch was also of very long duration, and again falls into three shorter periods, during which three great systems of strata were deposited, namely, first, the Devonian system, or the old red sandstone; upon that, the Carboniferous, or coal system; and upon this, the Permian system. The average thickness of these three systems taken together may amount to about 42,000 feet, from which we may infer the immense length of time requisite for their formation.

The Devonian and Permian formations are especially rich in remains of fishes, of primæval fish as well as enamelled fish (Ganoids), but the bony fish (Teleostei) are absent from the strata of the primary epoch. In coal are found the most ancient remains of animals living on land, both of articulated animals (spiders and insects) as well as of vertebrate animals (amphibious animals, like newts and frogs). In the Permian system there occur, in addition to the amphibious animals, the more highly-developed reptiles, and, indeed, forms nearly related to our lizards (Proterosaurus, etc.). But, nevertheless, we may call the primary epoch that of Fishes, because these few amphibious animals and reptiles are insignificant in comparison with the immense mass of palæozoic fishes. Just as Fishes predominate over the other vertebrate animals, so Ferns, or Filices, predominate among the plants of this epoch, and, in fact, real ferns and tree ferns (leafed ferns, or Phylopteridæ), as well as bamboo ferns (Calamophytæ) and scaled ferns (Lepidophytæ). These ferns, which grew on land, formed the chief part of the dense palæolithic island forests, the fossil remains of which are preserved to us in the enormously large strata of coal of the Carboniferous system, and in the smaller strata of coal of the Devonian and Permian systems. We are thus justified in calling the primary epoch either the era of Ferns or that of Fishes.

The third great division of the palæontological history of development is formed by the secondary epoch, or the era of Pine Forests, which is also called the mesolithic or mesozoic epoch. It extends from the end of the Permian system to the end of the Chalk formation, and is again divided into three great periods. The stratified systems deposited during this period are, first and lowest, the Triassic system, in the middle the Jura system, and at the top the Cretaceous system. The average thickness of these three systems taken together is much less than that of the primary group, and amounts as a whole only to about 15,000 feet. The secondary epoch can accordingly in all probability not have been half so long as the primary epoch.

Just as Fishes prevailed in the primary epoch, Reptiles predominated in the secondary epoch over all other vertebrate animals. It is true that during this period the first birds and mammals originated; at that time, also, there existed important amphibious animals, especially the gigantic Labyrinthodonts, in the sea the wonderful sea-dragons, or Halisaurii, swam about, and the first fish with bones were associated with the many primæval fishes (Sharks) and enamelled fish (Ganoids) of the earlier times; but the very variously developed kinds of reptiles formed the predominating and characteristic class of vertebrate animals of the secondary epoch. Besides those reptiles which were very nearly related to the present living lizards, crocodiles, and turtles, there were, during the mesolithic period, swarms of grotesquely shaped dragons. The remarkable flying lizards, or Pterosaurii, and the colossal land-dragons, or Dinosaurii, of the secondary epoch, are peculiar, as they occur neither in the preceding nor in the succeeding epochs. The secondary epoch may be called the era of Reptiles; but on the other hand, it may also be called the era of Pine Forests, or more accurately, of the Gymnosperms, that is, the epoch of plants having naked seeds. For this group of plants, especially as represented by the two important classes—the pines, or Coniferæ, and the palm-ferns, or Cycadeæ—during the secondary epoch constituted a predominant part of the forests. But towards the end of the epoch (in the Chalk period) the plants of the pine tribe gave place to the leaf-bearing forests which then developed for the first time.

SURVEY

Of the Palæontological Periods, or of the Greater Divisions of the Organic History of the Earth.

I. First Epoch: Archilithic Era.   Primordial Epoch.
(Era of Skull-less Animals and Forests of Tangles.)
1. Older Primordial Period or Laurentian Period.
2. Middle Primordial Period " Cambrian Period.
3. Later Primordial Period " Silurian Period.
II. Second Epoch: Palæolithic Era.   Primary Epoch.
(Era of Fish and Fern Forests.)
4. Older Primary Period or Devonian Period.
5. Mid Primary Period " Coal Period.
6. Later Primary Period " Permian Period.
III. Third Epoch: Mesolithic Era.   Secondary Epoch.
(Era of Reptiles and Pine Forests.)
7. Older Secondary Period or Trias Period.
8. Middle Secondary Period " Jura Period.
9. Later Secondary Period " Chalk Period.
IV. Fourth Epoch: Cænolithic Era.   Tertiary Epoch.
(Era of Mammals and Leaf Forests.)
10. Older Tertiary Period or Eocene Period.
11. Newer Tertiary Period " Miocene Period.
12. Recent Tertiary Period " Pliocene Period.
V. Fifth Epoch: Anthropolithic Era.   Quaternary Epoch.
(Era of Man and Cultivated Forests.)
13. Older Quaternary Period or Ice or Glacial Period.

14. Newer Quaternary Period " Post Glacial Period.
15. Recent Quaternary Period " Period of Culture.
(The Period of Culture is the Historical Period, or the Period of Tradition.)
The fourth main division of the organic history of the earth, the tertiary epoch, or era of Leafed Forests, is much shorter and less peculiar than the three first epochs. This epoch, which is also called the cænolithic or cænozoic epoch, extended from the end of the cretaceous system to the end of the pliocene system. The strata deposited during it amount only to a thickness of about 3,000 feet, and consequently are much inferior to the three first great groups. The three systems also into which the tertiary period is subdivided are very difficult to distinguish from one another. The oldest of them is called eocene, or old tertiary; the newer miocene, or mid tertiary; and the last is the pliocene, or later tertiary system.

The whole population of the tertiary epoch approaches much nearer, on the whole as well as in detail, to that of the present time than is the case in the preceding epochs. From this time the class of Mammals greatly predominates over all other vertebrate animals. In like manner, in the vegetable kingdom, the group—so rich in forms—of the Angiosperms, or plants with covered seeds, predominates, and its leafy forests constitute the characteristic feature of the tertiary epoch. The group of the Angiosperms consists of the two classes of single-seed-lobed plants, or Monocotyledons, and the double-seed-lobed plants, or Dicotyledons. The Angiosperms of both classes had, it is true, made their appearance in the Cretaceous period, and mammals had already occurred in the Jurassic period, and even in the Triassic period; but both groups, the mammals and the plants with enclosed seeds, did not attain their peculiar development and supremacy until the tertiary epoch, so that it may justly be called after them.

The fifth and last main division of the organic history of the earth is the quaternary epoch, or era of Civilization, which in comparison with the length of the four other epochs almost vanishes into nothing, though with a comical conceit we usually call its record the "history of the world." As the period is characterized by the development of Man and his Culture, which has influenced the organic world more powerfully and with greater transforming effect than have all previous conditions, it may also be called the era of Man, the anthropolithic or anthropozoic period. It might also be called the era of Cultivated Forests, or Gardens, because even at the lowest stage of human civilization man's influence is already perceptible in the utilization of forests and their products, and therefore also in the physiognomy of the landscape. The commencement of this era, which extends down to the present time, is geologically bounded by the end of the pliocene stratification.

The neptunic strata which have been deposited during the comparatively short quaternary epoch are very different in different parts of the earth, but they are mostly of very slight thickness. They are reduced to two "systems," the older of which is designated the diluvial, or pleistocene, and the later the alluvial, or recent. The diluvial system is again divided into two "formations," the older glacial and the more recent post glacial formations. For during the older diluvial period there occurred that extremely remarkable decrease of the temperature of the earth which led to an extensive glaciation of the temperate zones. The great importance which this "ice" or "glacial period" has exercised on the geographical and topographical distribution of organisms has already been explained in the preceding chapter (vol. i. p. 365). But the post glacial period, or the more recent diluvial period, during which the temperature again increased and the ice retreated towards the poles, was also highly important in regard to the present state of chorological relations.

The biological characteristic of the quaternary epoch lies essentially in the development and dispersion of the human organism and his culture. Man has acted with a greater transforming, destructive, and modifying influence upon the animal and vegetable population of the earth than any other organism. For this reason, and not because we assign to man a privileged exceptional position in nature in other matters, we may with full justice designate the development of man and his civilization as the beginning of a special and last main division of the organic history of the earth. It is probable indeed that the corporeal development of primæval man out of man-like apes took place as far back as the earlier pliocene period, perhaps even in the miocene tertiary period. But the actual development of human speech, which we look upon as the most powerful agency in the development of the peculiar characteristics of man and his dominion over other organisms, probably belongs to that period which on geological grounds is distinguished from the preceding pliocene period as the pleistocene or diluvial. In fact the time which has elapsed from the development of human speech down to the present day, though it may comprise many thousands and perhaps hundreds of thousands of years, almost vanishes into nothing as compared with the immeasurable length of the periods which have passed from the beginning of organic life on the earth down to the origin of the human race.

The tabular view given on page 15 shows the succession of the palæontological rock-groups, systems, and formations, that is, the larger and smaller neptunic groups of strata, which contain petrifactions, from the uppermost, or Alluvial, down to the lowest, or Laurentian, deposits. The table on page 14 presents the historical division of the corresponding eras of the larger and smaller palæontological periods, and in a reversed succession, from the most ancient Laurentian up to the most recent Quaternary period.

Many attempts have been made to make an approximate calculation of the number of thousands of years constituting these periods. The thickness of the strata has been compared, which, according to experience, is deposited during a century, and which amounts only to some few lines or inches, with the whole thickness of the stratified masses of rock, the succession of which we have just surveyed. This thickness, on the whole, may on an average amount to about 130,000 feet; of these 70,000 belong to the primordial, or archilithic; 42,000 to the primary, or palæolithic; 15,000 to the secondary, or mesolithic; and finally only 3,000 to the tertiary, or cænolithic group. The very small and scarcely appreciable thickness of the quaternary, or anthropolithic deposit cannot here come into consideration at all. On an average, it may at most be computed as from 500 to 700 feet. But it is self evident that all these measurements have only an average and approximate value, and are meant to give only a rough survey of the relative proportion of the systems of strata and of the spaces of time corresponding with them.

Now, if we divide the whole period of the organic history of the earth— that is, from the beginning of life on the earth down to the present day— into a hundred equal parts, and if then, corresponding to the thickness of the systems of strata, we calculate the relative duration of the time of the five main divisions or periods according to percentages, we obtain the following result:—

I. Archilithic, or primordial period 53.6
II. Palæolithic, or primary period 32.1
III. Mesolithic, or secondary period 11.5
IV. Cænolithic, or tertiary period 2.3
V. Anthropolithic, or quaternary period 0.5

———

Total ... 100.0

According to this, the length of the archilithic period, during which no land-living animals or plants as yet existed, amounts to more than one half, more than 53 per cent.; on the other hand the length of the anthropolithic era, during which man has existed, amounts to scarcely one-half per cent. of the whole length of the organic history of the earth. It is, however, quite impossible to calculate the length of these periods, even approximately, by years.

The thickness of the strata of mud at present deposited during a century, and which has been used as a basis for this calculation, is of course quite different in different parts of the earth under the different conditions in which these deposits take place. It is very slight at the bottom of the deep sea, in the beds of broad rivers with a short course, and in inland seas which receive very scanty supplies of water. It is comparatively great on the sea-shores exposed to strong breakers, at the estuaries of large rivers with long courses, and in inland seas with copious supplies of water. At the mouth of the Mississippi, which carries with it a considerable amount of mud, in the course of 100,000 years about 600 feet would be deposited. At the bottom of the open sea, far away from the coasts, during this long period only some few feet of mud would be deposited. Even on the sea-shores where a comparatively large quantity of mud is deposited the thickness of the strata formed during the course of a century may after all amount to no more than a few inches or lines when condensed into solid stone. In any case, however, all calculations based upon these comparisons are very unsafe, and we cannot even approximately conceive the enormous length of the periods which were requisite for the formation of the systems of neptunic strata. Here we can apply only relative, not absolute, measurements of time. Moreover, we should entirely err were we to consider the size of these systems of strata alone as the measure of the actual space of time which has elapsed during the earth's history. For the elevations and depressions of the earth's crust have perpetually alternated with one another, and the mineralogical and palæontological difference—which is perceived between each two succeeding systems of strata, and between each two of their formations at any particular spot—corresponds in all probability with a considerable intermediate space of many thousands of years, during which that particular part of the earth's crust was raised above the water. It was only after the lapse of this intermediate period, when a new depression again laid the part in question under water, that there occurred a new deposit of earth. As, in the mean time, the inorganic and organic conditions on this part had undergone a considerable transformation, the newly-formed layer of mud was necessarily composed of different earthy constituents and enclosed different petrifactions.

IV. Tertiary Group of Strata, 3,000 feet. Eocene, Miocene, Pliocene.

III. Mesolithic Group of Strata.

Deposits of the Secondary Epoch,
about 15,000 feet. IX. Chalk System.

VIII. Jura System.

VII. Trias System.

II. Palæolithic Group of Strata.

Deposits of the Primary Epoch,
about 42,000 feet. VI. Permian System.

V. Coal System.

IV. Devonian System.

I. Archilithic Group of Strata.

Deposits of the Primordial Epoch,
about 70,000 feet. III. Silurian System,
about 22,000 feet.

II. Cambrian System,
about18,000 feet.

I. Laurentian System
about 30,000 feet.

The striking differences which so frequently occur between the petrifactions of two strata, lying one above another, are to be explained in a simple and easy manner by the supposition that the same part of the earth's surface has been exposed to repeated depressions and elevations. Such alternating elevations and depressions take place even now extensively, and are ascribed to the heaving of the fiery fluid nucleus against the rigid crust. Thus, for example, the coast of Sweden and a portion of the west coast of South America are constantly though slowly rising, while the coast of Holland and a portion of the east coast of South America are gradually sinking. The rising as well as the sinking takes place very slowly, and in the course of a century sometimes only amounts to some few lines, sometimes to a few inches, or at most a few feet. But if this action continues uninterruptedly throughout hundreds of thousands of years it is capable of forming the highest mountains.

It is evident that elevations and depressions, such as now can be measured in these places, have uninterruptedly alternated one with another in different places during the whole course of the organic history of the earth. This may be inferred with certainty from the geographical distribution of organisms. (Compare vol. i. p. 350.) But to form a judgment of our palæontological records of creation it is extremely important to show that permanent strata can only be deposited during a slow sinking of the ground under water, but not during its continued rising. When the ground slowly sinks more and more below the level of the sea, the deposited layers of mud get into continually deeper and quieter water, where they can become condensed into stone undisturbed. But when, on the other hand, the ground slowly rises, the newly-deposited layers of mud, which enclose the remains of plants and animals, again immediately come within the reach of the play of the waves, and are soon worn away by the force of the breakers, together with the organic remains which they on close. For this simple but very important reason, therefore, abundant layers, in which organic remains are preserved, can only be deposited during a continuous sinking of the ground. When any two different formations or strata, lying one above the other, correspond with two different periods of depression, we must assume a long period of rising between them, of which period we know nothing, because no fossil remains of the then living animals and plants could be preserved. It is evident, however, that those periods of elevation, which have passed without leaving any trace behind them, deserve a no less careful consideration than the greater or less alternating periods of depression, of whose organic population we can form an approximate idea from the strata containing petrifactions. Probably the former were not of shorter duration than the latter.

From this alone it is apparent how imperfect our records must necessarily be, and all the more so since it can be theoretically proved that the variety of animal and vegetable life must have increased greatly during those very periods of elevation. For as new tracts of land are raised above the water, new islands are formed. Every new island, however, is a new centre of creation, because the animals and plants accidentally cast ashore there, find in the new territory, in the struggle for life, abundant opportunity of developing themselves peculiarly, and of forming new species. The formation of new species has evidently taken place pre-eminently during these intermediate periods, of which, unfortunately, no petrifactions could be preserved, whereas, on the contrary, during the slow sinking of the ground there was more chance of numerous species dying out, and of a retrogression into fewer specific forms. The intermediate forms between the old and the newly-forming species must also have lived during the periods of elevation, and consequently could likewise leave no fossil remains.

In addition to the great and deplorable gaps in the palæontological records of creation—which are caused by the periods of elevation—there are, unfortunately, many other circumstances which immensely diminish their value. I must mention here especially the metamorphic state of the most ancient formations, of those strata which contain the remains of the most ancient flora and fauna, the original forms of all subsequent organisms, and which, therefore, would be of especial interest. It is just these rocks—and, indeed, the greater part of the primordial, or archilithic strata, almost the whole of the Laurentian, and a large part of the Cambrian systems—which no longer contain any recognizable remains, and for the simple reason that these strata have been subsequently changed or metamorphosed by the influence of the fiery fluid interior of the earth. These deepest neptunic strata of the crust have been completely changed from their original condition by the heat of the glowing nucleus of the earth, and have assumed a crystalline state. In this process, however, the form of the organic remains enclosed in them has been entirely destroyed. It has been preserved only here and there by a happy chance, as in the case of the most ancient petrifactions known, the Eozoon canadense, from the lowest Laurentian strata. However, from the layers of crystalline charcoal (graphite) and crystalline limestone (marble), which are found deposited in the metamorphic rocks, we may with certainty conclude that petrified animal and vegetable remains existed in them in earlier times.

Our record of creation is also extremely imperfect from the circumstance that only a small portion of the earth's surface has been accurately investigated by geologists, namely, England, Germany, and France. But we know very little of the other parts of Europe, of Russia, Spain, Italy, and Turkey. In the whole of Europe, only some few parts of the earth's crust have been laid open, by far the largest portion of it is unknown to us. The same applies to North America and to the East Indies. There some few tracts have been investigated; but of the larger portion of Asia, the most extensive of all continents, we know almost nothing; of Africa nothing, excepting the Cape of Good Hope and the shores of the Mediterranean; of Australia almost nothing; and of South America but very little. It is clear, therefore, that only quite a small portion, perhaps scarcely the thousandth part of the whole surface of the earth, has been palæontologically investigated. We may therefore reasonably hope, when more extensive geological investigations are made, which are greatly assisted by the constructions of railroads and mines, to find a great number of other important petrifactions. A hint that this will be the case is given by the remarkable petrifactions found in those parts of Africa and Asia which have been minutely investigated,—the Cape districts and the Himalaya mountains. A series of entirely new and very peculiar animal forms have become known to us from the rocks of these localities. But we must bear in mind that the vast bottom of the existing oceans is at the present time quite inaccessible to palæontological investigations, and that the greater part of the petrifactions which have lain there from primæval times will either never be known to us, or at best only after the course of many thousands of years, when the present bottom of the ocean shall have become accessible by gradual elevation. If we call to mind the fact that three-fifths of the whole surface of the earth consists of water, and only two-fifths of land, it becomes plain that on this account the palæontological record must always present an immense gap.

But, in addition to these, there exists another series of difficulties in the way of palæontology which arises from the nature of the organisms themselves. In the first place, as a rule only the hard and solid parts of organisms can fall to the bottom of the sea or of fresh waters, and be there enclosed in the mud and petrified. Hence it is only the bones and teeth of vertebrate animals, the calcareous shells of molluscs, the chitinous skeletons of articulated animals, the calcareous skeletons of star-fishes and corals, and the woody and solid parts of plants, that are capable of being petrified. But soft and delicate parts, which constitute by far the greater portion of the bodies of most organisms, are very rarely deposited in the mud under circumstances favourable to their becoming petrified, or distinctly impressing their external form upon the hardening mud. Now, it must be borne in mind that large classes of organisms, as for example the Medusæ, the naked molluscs without shells, a large portion of the articulated animals, almost all worms, and even the lowest vertebrate animals, possess no firm and hard parts capable of being petrified. In like manner the most important parts of plants, such as the flowers, are for the most part so soft and tender that they cannot be preserved in a recognizable form. We therefore cannot expect to find any petrified remains of these important organisms. Moreover, all organisms at an early stage of life are so soft and tender that they are quite incapable of being petrified. Consequently all the petrifactions found in the neptunic stratifications of the earth's crust comprise altogether but a very few forms, and of these for the most part only isolated fragments.

We must next bear in mind that the dead bodies of the inhabitants of the sea are much more likely to be preserved and petrified in the deposits of mud than those of the inhabitants of fresh water and of the land. Organisms living on land can, as a rule, become petrified only when their corpses fall accidentally into the water and are buried at the bottom in the hardening layers of mud. But this event depends upon very many conditions. We cannot therefore be astonished that by far the majority of petrifactions belong to organisms which have lived in the sea, and that of the inhabitants of the land proportionately only very few are preserved in a fossil state. How many contingencies come into play here we may infer from the single fact that of many fossil mammals, in fact of all the mammals of the secondary, or mesozoic epoch, nothing is known except the lower jawbone. This bone is in the first place comparatively solid, and in the second place very easily separates itself from the dead body, which floats on the water. Whilst the body is driven away and dissolved by the water, the lower jawbone falls down to the bottom of the water and is there enclosed in the mud. This explains the remarkable fact that in a stratum of limestone of the Jurassic system near Oxford, in the slates of Stonesfield, as yet only the lower jawbones of numerous pouched animals (Marsupials) have been found. They are the most ancient mammals known, and of the whole of the rest of their bodies not a single bone exists. The opponents of the theory of development, according to their usual logic, would from this fact be obliged to draw the conclusion that the lower jawbone was the only bone in the body of those animals.

Footprints are very instructive when we attempt to estimate the many accidents which so arbitrarily influence our knowledge of fossils; they are found in great numbers in different extensive layers of sandstone; for example, in the red sandstone of Connecticut, in North America. These footprints were evidently made by vertebrate animals, probably by reptiles, of whose bodies not the slightest trace has been preserved.[1] The impressions which their feet have left on the mud alone betray the former existence of these otherwise unknown animals.

The accidents which, besides these, determine the limits of our palæontological knowledge, may be inferred from the fact that we know of only one or two specimens of very many important petrifactions. It is not ten years since we became acquainted with the imperfect impression of a bird in the Jurassic or Oolitic system, the knowledge of which has been of the very greatest importance for the phylogeny of the whole class of birds. All birds previously known presented a very uniformly organized group, and showed no striking transitional forms to other vertebrate classes, not even to the nearly related reptiles. But that fossil bird from the Jura possessed not an ordinary bird's tail, but a lizard's tail, and thus confirmed what had been conjectured upon other grounds, namely, the derivation of birds from lizards. This single fossil has thus essentially extended not only our knowledge of the age of the class of birds, but also of their blood relationship to reptiles. In like manner our knowledge of other animal groups has been often essentially modified by the accidental discovery of a single fossil. The palæontological records must necessarily be exceedingly imperfect, because we know of so very few examples, or only mere fragments of very many important fossils.

Another and very sensible gap in these records is caused by the circumstance that the intermediate forms which connect the different species have, as a rule, not been preserved, and for the simple reason that (according to the principle of divergence of character) they were less favoured in the struggle for life than the most divergent varieties, which had developed out of one and the same original form. The intermediate links have, on the whole, always died out rapidly, and have but rarely been preserved as fossils. On the other hand, the most divergent forms were able to maintain themselves in life for a longer period as independent species, to propagate more numerously, and consequently to be more readily petrified. But this does not exclude the fact that in some cases the connecting intermediate forms of the species have been preserved so perfectly petrified, that even now they cause the greatest perplexity and occasion endless disputes among systematic palæontologists about the arbitrary limits of species.

An excellent example of this is furnished by the celebrated and very variable fresh-water snail from the Stuben Valley, near Steinheim, in Würtemburg, which has been described sometimes as Paludina, sometimes as Valvata, and sometimes as Planorbis multiformis. The snow-white shells of these small snails constitute more than half of the mass of the tertiary limestone hills, and in this one locality show such an astonishing variety of forms, that the most divergent extremes might be referred to at least twenty entirely different species. But all these extreme forms are united by such innumerable intermediate forms, and they lie so regularly above and beside one another, that Hilgendorf was able, in the clearest manner, to unravel the pedigree of the whole group of forms. In like manner, among very many other fossil species (for example, many ammonites, terebratulæ, sea urchins, lily encrinites, etc.) there are such masses of connecting intermediate forms, that they reduce the "dealers in fossil species" to despair.

When we weigh all the circumstances here mentioned, the number of which might easily be increased, it does not appear astonishing that the natural accounts or records of creation formed by petrifactions are extremely defective and incomplete. But nevertheless, the petrifactions actually discovered are of the greatest value. Their significance is of no less importance to the natural history of creation than the celebrated inscription on the Rosetta stone, and the decree of Canopus, are to the history of nations—to archæology and philology. Just as it has become possible by means of these two most ancient inscriptions to reconstruct the history of ancient Egypt, and to decipher all hieroglyphic writings, so in many cases a few bones of an animal, or imperfect impressions of a lower animal or vegetable form, are sufficient for us to gain the most important starting-points in the history of the whole group, and in the search after their pedigree. A couple of small back teeth, which have been found in the Keuper formation of the Trias, have of themselves alone furnished a sure proof that mammals existed even in the Triassic period.

Of the incompleteness of the geological accounts of creation, Darwin, agreeing with Lyell, the greatest of all recent geologists, says:—

"I look at the geological record as a history of the world imperfectly kept, and written in a changing dialect; of this history we possess the last volume alone, relating only to two or three countries. Of this volume, only here and there a short chapter has been preserved; and of each page, only here and there a few lines. Each word of the slowly-changing language, more or less different in the successive chapters, may represent the forms of life which are entombed in our consecutive formations, and which falsely appear to us to have been abruptly introduced. On this view, the difficulties above discussed are greatly diminished, or even disappear."—Origin of Species, 6th Edition, p. 289.

If we bear in mind the exceeding incompleteness of palæontological records, we shall not be surprised that we are still dependent upon so many uncertain hypotheses when actually endeavouring to sketch the pedigree of the different organic groups. However, we fortunately possess, besides fossils, other records of the history of the origin of organisms, which in many cases are of no less value, nay, in several cases are of much greater value, than fossils. By far the most important of these other records of creation is, without doubt, ontogeny, that is, the history of the development of the organic individual (embryology and metamorphology). It briefly repeats in great and marked features the series of forms which the ancestors of the respective individuals have passed through from the beginning of their tribe. We have designated the palæontological history of the development of the ancestors of a living form as the history of a tribe, or phylogeny, and we may therefore thus enunciate this exceedingly important biogenetic fundamental principle: "Ontogeny is a short and quick repetition, or recapitulation, of Phylogeny, determined by the laws of Inheritance and Adaptation." As every animal and every plant from the beginning of its individual existence passes through a series of different forms, it indicates in rapid succession and in general outlines the long and slowly changing series of states of form which its progenitors have passed through from the most ancient times. (Gen. Morph. ii. 6, 110, 300.)

It is true that the sketch which the ontogeny of organisms gives us of their phylogeny is in most cases more or less obscured, and all the more so the more Adaptation, in the course of time, has predominated over Inheritance, and the more powerfully the law of abbreviated inheritance, and the law of correlative adaptation, have exerted their influence. However, this does not lessen the great value which the actual and faithfully preserved features of that sketch possess. Ontogeny is of the most inestimable value for the knowledge of the earliest palæontological conditions of development, just because no petrified remains of the most ancient conditions of the development of tribes and classes have been preserved. These, indeed, could not have been preserved on account of the soft and tender nature of their bodies. No petrifactions could inform us of the fundamental and important fact which ontogeny reveals to us, that the most ancient common ancestors of all the different animal and vegetable species were quite simple cells like the egg-cell. No petrifaction could prove to us the immensely important fact, established by ontogeny, that the simple increase, the formation of cell-aggregates and the differentiation of those cells, produced the infinitely manifold forms of multicellular organisms. Thus ontogeny helps us over many and large gaps in palæontology.

To the invaluable records of creation furnished by palæontology and ontogeny are added the no less important evidences for the blood relationship of organisms furnished by comparative anatomy. When organisms, externally very different, nearly agree in their internal structure, one may with certainty conclude that the agreement has its foundation in Inheritance, the dissimilarity its foundation in Adaptation. Compare, for example, the hands and fore paws of the nine different animals which are represented on Plate IV., in which the bony skeleton in the interior of the hand and of the five fingers is visible. Everywhere we find, though the external forms are most different, the same bones, and among them the same number, position, and connection. It will perhaps appear very natural that the hand of man (Fig. 1) differs very little from that of the gorilla (Fig. 2) and of the orang-outang (Fig. 3), his nearest relations. But it will be more surprising if the fore feet of the dog also (Fig. 4), as well as the breast-fin (the hand) of the seal (Fig. 5), and of the dolphin (Fig. 6), show essentially the same structure. And it will appear still more wonderful that even the wing of the bat (Fig. 7), the shovel-feet of the mole (Fig. 8), and the fore feet of the duck-bill (Ornithorhynchus) (Fig. 9), the most imperfect of all mammals, is composed of entirely the same bones, only their size and form being variously changed. Their number, the manner of their arrangement and connection has remained the same. (Compare also the explanation of Plate IV., in the Appendix.) It is quite inconceivable that any other cause, except the common inheritance of the part in question from common ancestors, could have occasioned this wonderful homology or similarity in the essential inner structure with such different external forms. Now, if we go down further in the system below the mammals, and find that even the wings of birds, the fore feet of reptiles and amphibious animals, are composed of essentially the same bones as the arms of man and the fore legs of the other mammals, we can, from this circumstance alone, with perfect certainty, infer the common origin of all these vertebrate animals. Here, as in all other cases, the degree of the internal agreement in the form discloses to us the degree of blood relationship.

PEDIGREE AND HISTORY OF THE KINGDOM OF THE PROTISTA

By a careful comparison of the individual and the palæontological development, as also by the comparative anatomy of organisms, by the comparative examination of their fully developed structural characteristics, we arrive at the knowledge of the degrees of their different structural relationships. By this, however, we at the same time obtain an insight into their true blood relationship, which, according to the Theory of Descent, is the real reason of the structural relationship. Hence by collecting, comparing, and employing the empirical results of embryology, palæontology, and anatomy for supplementing each other, we arrive at an approximate knowledge of "the Natural System," which, according to our views, is the pedigree of organisms. It is true that our human knowledge, in all things fragmentary, is especially so in this case, on account of the extreme incompleteness and defectiveness of the records of creation. However, we must not allow this to discourage us, or to deter us from undertaking this highest problem of biology. Let us rather see how far it may even now be possible, in spite of the imperfect state of our embryological, palæontological, and anatomical knowledge, to establish a probable scheme of the genealogical relationships of organisms.

Darwin in his book gives us no answer to these special questions of the Theory of Descent; at the conclusion he only expresses his conjecture "that animals have descended from at most only four or five progenitors, and plants from an equal or less number." But as these few aboriginal forms still show traces of relationship, and as the animal and vegetable kingdoms are connected by intermediate transitional forms, he arrives afterwards at the opinion "that probably all the organic beings which have ever lived on the earth have descended from some one primordial form, into which life was

25

first breathed by the Creator." Like Darwin, all other adherents of the Theory of Descent have only treated it in a general way, and not made the attempt to carry it out specially, and to treat the "Natural System" actually as the pedigree of organisms. If, therefore, we venture upon this difficult undertaking, we must take up independent ground.

Four years ago I set up a number of hypothetical genealogies for the larger groups of organisms in the systematic introduction to my General History of Development (Gen. Morph. vol. ii.), and thereby, in fact, made the first attempt actually to construct the pedigrees of organisms in the manner required by the theory of development. I was quite conscious of the extreme difficulty of the task, and as I undertook it in spite of all discouraging obstacles, I claim no more than the merit of having made the first attempt and given a stimulus for other and better attempts. Probably most zoologists and botanists were but little satisfied with this beginning, and least so in reference to the special domain in which each one is specially at work. However, it is certainly in this case much easier to blame than to produce something better, and what best proves the immense difficulty of this infinitely complicated task is the fact that no naturalist has as yet supplied the place of my pedigrees by better ones. But, like all other scientific hypotheses which serve to explain facts, my genealogical hypotheses may claim to be taken into consideration until they are replaced by better ones.

I hope that this replacement will very soon take place; and I wish for nothing more than that my first attempt may induce very many naturalists to establish more accurate pedigrees for the individual groups, at least in the special domain of the animal and vegetable kingdom which happens to be well known to one or other of them. By numerous attempts of this kind our genealogical knowledge, in the course of time, will slowly advance and approach more towards perfection, although it can with certainty be foreseen that we shall never arrive at a complete pedigree. We lack, and shall ever lack, the indispensable palæontological foundations. The most ancient records will ever remain sealed to us, for reasons which have been previously mentioned. The most ancient organisms which arose by spontaneous generation—the original parents of all subsequent organisms—must necessarily be supposed to have been Monera—simple, soft, albuminous lumps, without structure, without any definite forms, and entirely without any hard and formed parts. They and their next offspring were consequently not in any way capable of being preserved in a petrified condition. But we also lack, for reasons discussed in detail in the preceding chapter, by far the greater portion of the innumerable palæontological documents, which are really requisite for a safe reconstruction of the history of animal tribes, or phylogeny, and for the true knowledge of the pedigree of organisms. If we, therefore, in spite of this, venture to undertake their hypothetical construction, we must chiefly depend for guidance on the two other series of records which most essentially supplement the palæontological archives. These are ontogeny and comparative anatomy.

If thoughtfully and carefully we consult these most valuable records, we at once perceive what is exceedingly significant, namely, that by far the greater number of organisms, especially all higher animals and plants, are composed of a great number of cells, and that they originate out of an egg, and that this egg, in animals as well as in plants, is a single, perfectly simple cell—a little lump of albuminous constitution, in which another albuminous corpuscle, the cell-kernel, is enclosed. This cell containing its kernel grows and becomes enlarged. By division it forms an accumulation of cells, and out of these, by division of labour (as has previously been described), there arise the numberless different forms which are presented to us in the fully developed animal and vegetable species. This immensely important process—which we may follow step by step, with our own eyes, any day in the embryological development of any animal or vegetable individual, and which as a rule is by no means considered with the reverence it deserves— informs us more surely and completely than all petrifactions could do as to the original palæontological development of all many-celled organisms, that is, of all higher animals and plants. For as ontogeny, or the embryological development of every single individual, is essentially only a recapitulation of phylogeny, or the palæontological development of its chain of ancestors, we may at once, with full assurance, draw the simple and important conclusion, that all many-celled animals and plants were originally derived from single-celled organisms. The primæval ancestors of man, as well as of all other animals, and of all plants composed of many cells, were simple cells living isolated. This invaluable secret of the organic pedigree is revealed to us with infallible certainty by the egg of animals, and by the true egg-cell of plants. When the opponents of the Theory of Descent assert it to be miraculous and inconceivable that an exceedingly complicated many-celled organism could, in the course of time, have proceeded from a simple single-celled organism, we at once reply that we may see this incredible miracle at any moment, and follow it with our own eyes. For the embryology of animals and plants visibly presents to our eyes in the shortest space of time the same process as that which has taken place in the origin of the whole tribe during the course of enormous periods of time.

Upon the ground of embryological records, therefore, we can with full assurance maintain that all many-celled, as well as single-celled, organisms are originally descended from simple cells; connected with this, of course, is the conclusion that the most ancient root of the animal and vegetable kingdom was common to both. For the different primæval "original cells" out of which the few different main groups or tribes have developed, only acquired their differences after a time, and were descended from a common "primæval cell." But where did those few "original cells," or the one primæval cell, come from? For the answer to this fundamental genealogical question we must return to the theory of plastids and the hypothesis of spontaneous generation which we have already discussed (vol. i. p. 327).

As was then shown, we cannot imagine cells to have arisen by spontaneous generation, but only Monera, those primæval creatures of the simplest kind conceivable, like the still living Protamœbæ), Protomyxæ, etc. (vol. i. p. 1186, Fig. 1). Only such corpuscules of mucus without component parts—whose whole albuminous body is as homogeneous in itself as an inorganic crystal, but which nevertheless fulfills the two organic fundamental functions of nutrition and propagation—could have directly arisen out of inorganic matter by autogeny at the beginning (we may suppose) of the Laurentian period. While some Monera remained at the original simple stage of formation, others gradually developed into cells by the inner kernel of the albuminous mass becoming separated from the external cell-substance. In others, by differentiation of the outermost layer of the cell-substance, an external covering (membrane, or skin) was formed round simple cytods (without kernel), as well as round naked cells (containing a kernel). By these two processes of separation in the simple primæval mucus of the Moneron body, by the formation of a kernel in the interior and a covering on the outer surface of the mass of plasma, there arose out of the original most simple cytods, or Monera, those four different species of plastids, or individuals, of the first order, from which, by differentiation and combination, all other organisms could afterwards develop themselves. (Compare vol. i. p. 347.)

The question now forces itself upon us, Are all organic cytods and cells, and consequently also those "original cells" which we previously considered to be the primary parents of the few great main groups of the animal and vegetable kingdoms, descended from a single original form of Moneron, or were there several different organic primary forms, each traceable to a peculiar independent species of Moneron which originated by spontaneous generation? In other words, Is the whole organic world of a common origin, or does it owe its origin to several acts of spontaneous generation? This fundamental question of genealogy seems at first sight to be of exceeding importance. But on a more accurate examination, we shall soon see that this is not the case, and that it is in reality a matter of very subordinate importance.

Let us now pass on to examine and clearly limit our conception of an organic tribe. By tribe, or phylum, we understand all those organisms of whose blood relationship and descent from a common primary form there can be no doubt, or whose relationship, at least, is most probable from anatomical reasons, as well as from reasons founded on historical development. Our tribes, or phyla, according to this idea, essentially coincide with those few "great classes," or "main classes," of which Darwin also thinks that each contains only organisms related by blood, and of which, both in the animal and in the vegetable kingdoms, he only assumes either four or five. In the animal kingdom these tribes would essentially coincide with those four, five, or six main divisions which zoologists, since Bär and Cuvier, have distinguished as "main forms, general plans, branches, or sub-kingdoms" of the animal kingdom. (Compare vol. i. p. 53.) Bär and Cuvier distinguished only four of them, namely:—1. The vertebrate animals (Vertebrata); 2. The articulated animals (Articulata); 3. The molluscous animals (Mollusca); and 4. The radiated animals (Radiata). At present six are generally distinguished, since the tribe of the articulated animals is divided into two tribes, those possessing articulated feet (Arthropoda), and the worms (Vermes); and in like manner the tribe of radiated animals is subdivided into the two tribes of the star animals (Echinodermata) and the animal-plants (Zoophyta). Within each of these six tribes, all the included animals, in spite of great variety in external form and inner structure, nevertheless possess such numerous and important characteristics in common, that there can be no doubt of their blood relationship. The same applies also to the six great main classes which modern botany distinguishes in the vegetable kingdom, namely:—1. Flowering plants (Phanerogamia); 2. Ferns (Filicinæ); 3. Mosses (Muscinæ); 4. Lichens (Lichenes); 5. Fungi (Fungi); and 6. Water-weeds (Algæ). The last three groups, again, show such close relations to one another, that by the name of "Thallus plants" they may be contrasted with the three first main classes, and consequently the number of phyla, or main groups, of the vegetable kingdom may be reduced to the number of four. Mosses and ferns may likewise be comprised as "Prothallus plants" (Prothallophyta), and thereby the number of plant tribes reduced to three—Flowering plants, Prothallus plants, and Thallus plants.

Very important facts in the anatomy and the history of development, both in the animal and vegetable kingdoms, support the supposition that even these few main classes or tribes are connected at their roots, that is, that the lowest and most ancient primary forms of all three are related by blood to one another. Nay, by a further examination we are obliged to go still a step further, and to agree with Darwin's supposition, that even the two pedigrees of the animal and vegetable kingdom are connected at their lowest roots, and that the lowest and most ancient animals and plants are derived from a single common primary creature. According to our view, this common primæval organism can have been nothing but a Moneron which took its origin by spontaneous generation.

In the mean time we shall at all events be acting cautiously if we avoid this last step, and assume true blood relationship only within each tribe, or phylum, where it has been undeniably and surely established by facts in comparative anatomy, ontogeny, and phylogeny. But we may here point to the fact that two different fundamental forms of genealogical hypothesis are possible, and that all the different investigations of the Theory of Descent in relation to the origin of organic groups of forms will, in future, tend more and more in one or the other of these directions. The unitary, or monophyletic, hypothesis of descent will endeavour to trace the first origin of all individual groups of organisms, as well as their totality, to a single common species of Moneron which originated by spontaneous generation (vol. i. p. 343). The multiple, or polyphyletic, hypothesis of descent, on the other hand, will assume that several different species of Monera have arisen by spontaneous generation, and that these gave rise to several different main classes (tribes, or phyla) (vol. i. p. 348). The apparently great contrast between these two hypotheses is in reality of very little importance. For both the monophyletic and the polyphyletic hypothesis of descent must necessarily go back to the Monera as the most ancient root of the one or of the many organic tribes. But as the whole body of a Moneron consists only of a simple, formless mass, without component particles, made up of a single albuminous combination of carbon, it follows that the differences of the different Monera can only be of a chemical nature, and can only consist in a different atomic composition of that mucous albuminous combination. But these subtle and complicated differences of mixture of the infinitely manifold combinations of albumen are not appreciable by the rude and imperfect means of human observation and are, consequently, at present of no further interest to the task we have in hand.

The question of the monophyletic or polyphyletic origin will constantly recur within each individual tribe, where the origin of a smaller or of a larger group is discussed. In the vegetable kingdom, for example, some botanists will be inclined to derive all flowering plants from a single form of fern, while others will prefer the idea that several different groups of Phanerogama have sprung from several different groups of ferns. In like manner, in the animal kingdom, some zoologists will be more in favour of the supposition that all placental animals are derived from a single pouched animal; others will be more in favour of the opposite supposition, that several different groups of placental animals have proceeded from several different pouched animals. In regard to the human race itself, some will prefer to derive it from a single form of ape, while others will be more inclined to the idea that several different races of men have arisen, independently of one another, out of several different species of ape. Without here expressing our opinion in favour of either the one or the other conception, we must, nevertheless, remark that in general the monophyletic hypothesis of descent deserves to be preferred to the polyphyletic hypothesis of descent. In accordance with the chorological proposition of a single "centre of creation" or of a single primæval home for most species (which has already been discussed), we may be permitted to assume that the original form of every larger or smaller natural group only originated once in the course of time, and only in one part of the earth. We may safely assume this simple original root, that is, the monophyletic origin, in the case of all the more highly developed groups of the animal and vegetable kingdoms. (Compare vol. i. p. 353.) But it is very possible that the more complete Theory of Descent of the future will involve the polyphyletic origin of very many of the low and imperfect groups of the two organic kingdoms.

For these reasons I consider it best, in the mean time, to adopt the monophyletic hypothesis of descent both for the animal and for the vegetable kingdom. Accordingly, the above-mentioned six tribes, or phyla, of the animal kingdom must be connected at their lowest root, and likewise the three or six main classes, or phyla, of the vegetable kingdom must be traced to a common and most ancient original form. How the connection of these tribes is to be conceived I shall explain in the succeeding chapters. But before proceeding to this, we must occupy ourselves with a very remarkable group of organisms, which cannot without artificial constraint be assigned either to the pedigree of the vegetable or to that of the animal kingdom. These interesting and important organisms are the primary creatures, or Protista.

All organisms which we comprise under the name of Protista show in their external form, in their inner structure, and in all their vital phenomena, such a remarkable mixture of animal and vegetable properties, that they cannot with perfect justice be assigned either to the animal or to the vegetable kingdom; and for more than twenty years an endless and fruitless dispute has been carried on as to whether they are to be assigned to this or that kingdom. Most of Protista are so small that they can scarcely, if at all, be perceived with the naked eye. Hence the majority of them have only become known during the last fifty years, since by the help of the improved and general use of the microscope these minute organisms have been more frequently observed and more accurately examined. However, no sooner were they better known than endless disputes arose about their real nature and their position in the natural system of organisms. Many of these doubtful primary creatures botanists defined as animals, and zoologists as plants; neither of the two would own them. Others, again, were declared by botanists to be plants, and by zoologists to be animals; each claimed them. These contradictions are not altogether caused by our imperfect knowledge of the Protista, but in reality by their true nature. Indeed, most Protista present such a confused mixture of several animal and vegetable characteristics, that each investigator may arbitrarily assign them either to the animal or vegetable kingdom. Accordingly as he defines these two kingdoms, and as he looks upon this or that characteristic as determining the animal or vegetable nature, he will assign the individual classes of Protista in one case to the animal and in another to the vegetable kingdom. But this systematic difficulty has become an inextricable knot by the fact that all more recent investigations on the lowest organisms have completely effaced, or at least destroyed, the sharp boundary between the animal and vegetable kingdom which had hitherto existed, and to such a degree that its restoration is possible only by means of a completely artificial definition of the two kingdoms. But this definition could not be made so as to apply to many of the Protista.

For this and other reasons it is, in the mean time, best to exclude the doubtful beings from the animal as well as from the vegetable kingdom, and to comprise them in a third organic kingdom standing midway between the two others. This intermediate kingdom I have established as the Kingdom of the Primary Creatures (Protista), when discussing general anatomy in the first volume of my General Morphology, pp. 191-238. In my Monograph of the Monera,(15) I have recently treated of this kingdom, having somewhat changed its limits, and given it a more accurate definition. Of independent classes of the kingdom Protista, we may at present distinguish the following:—

1. The still living Monera; 2. The Amœboidea, or Protoplasts; 3. The Whip-swimmers, or Flagellata; 4. The Flimmer-balls, or Catallacta; 5. The Tram-weavers, or Labyrinthuleæ; 6. The Flint-cells, or Diatomeæ; 7. The Slime-moulds, or Myxomycetes; 8. The Ray-streamers, or Rhizopoda.

The most important groups at present distinguishable in these eight classes of Protista are named in the systematic table on p. 51. Probably the number of these Protista will be considerably increased in future days by the progressive investigations of the ontogeny of the simplest forms of life, which have only lately been carried on with any great zeal. With most of the classes named we have become intimately acquainted only during the last ten years. The exceedingly interesting Monera and Labyrinthuleæ, as also the Catallacta, were indeed discovered only a few years ago. It is probable also that very numerous groups of Protista have died out in earlier periods, without having left any fossil remains, owing to the very soft nature of their bodies. We might add to the Protista from the still living lowest groups of organisms—the Fungi; and in so doing should make a very large addition to its domain. Provisionally we shall leave them among plants, though many naturalists have separated them altogether from the vegetable kingdom.

The pedigree of the kingdom Protista is still enveloped in the greatest obscurity. The peculiar combination of animal and vegetable properties, the indifferent and uncertain character of their relations of forms and vital phenomena, together with a number of several very peculiar features which separate most of the subordinate classes sharply from the others, at present baffle every attempt distinctly to make out their blood relationships with one another, or with the lowest animals on the one hand, and with the lowest plants on the other hand. It is not improbable that the classes specified, and many other unknown classes of Protista, represent quite independent organic tribes, or phyla, each of which has independently developed from one, perhaps from various, Monera which have arisen by spontaneous generation. If we do not agree to this polyphyletic hypothesis of descent, and prefer the monophyletic hypothesis of the blood relationship of all organisms, we shall have to look upon the different classes of Protista as the lower small offshoots of the root, springing from the same simple Monera root, out of which arose the two mighty and many-branched pedigrees of the animal kingdom on the one hand, and of the vegetable kingdom on the other. (Compare pp. 74, 75.) Before I enter into this difficult question more accurately, it will be appropriate to premise something further as to the contents of the classes of Protista given on the next page, and their general natural history.

SYSTEMATIC SURVEY

Of the Larger and Smaller Groups of the Kingdom Protista

Classes of
the Protista
Kingdom. Systematic Name
of the Classes Orders of
Families of the
Classes. A name of a
Genus
as an example.

1. Moners   Monera
1. Gymnomonera Protogenes
2. Lepomonera Protomyxa
2. Protoplasts Amœboida
1. Gymnamœbæ Amœba
2. Leptamœbæ Arcella
3. Gregarinæ Monocystis
3. Whip-swimmers
Flagellata
1. Nudiflagellata Euglena
2. Cilioflagellata Peridinium
4. Flimmer-balls   Catallacta   1. Catallacta Magosphæra
5. Tram-weavers Labyrinthuleæ   1. Labyrinthuleæ Labyrinthula
6. Flint-cells Diatomea
1. Striata Navicula
2. Vittata Tabellaria
3. Areolata Coscinodiscus
7. Slime-moulds Myxomycetes

1. Physareæ Æthalium
2. Stemoniteæ Stemonitis
3. Trichiaceæ Arcyria
4. Lycogaleæ Reticularia
8. Ray-streamers or Rhizopods.
(Root-feet)
I. Acyttaria
1. Monothalamia Gromia
2. Polythalamia Nummulina
II. Heliozoa
1. Heliozoa Actinosphærium
III. Radiolaria
1. Monocyttaria Cyrtidosphæra
2. Polycyttaria Collosphæra

It will perhaps seem strange that I should here again begin with the remarkable Monera as the first class of the Protista kingdom, as I of course look upon them as the most ancient primary forms of all organisms without exception. Still, what are we otherwise to do with the still living Monera? We know nothing of their palæontological origin, we know nothing of any of their relations to lower animals or plants, and we know nothing of their possible capability of developing into higher organisms. The simple and homogeneous little lump of slime or mucus which constitutes their entire body (Fig. 8) is the most ancient and original form of animal as well as of vegetable plastids. Hence it would evidently be just as arbitrary and unreasonable to assign them to the animal as it would be to assign them to the vegetable kingdom. In any case we shall for the present be acting more cautiously and critically if we comprise the still living Monera—whose number and distribution is probably very great—as a special and independent class, contrasting them with the other classes of the kingdom Protista, as well as with the animal kingdom. Morphologically considered, the Monera—on account of the perfect homogeneity of the albuminous substance of their bodies, on account of their utter want of heterogeneous particles—are more closely connected with anorgana than with organisms, and evidently form the transition between the inorganic and organic world of bodies, as is necessitated by the hypothesis of spontaneous generation. I have described and given illustrations of the forms and vital phenomena of the still living Monera (Protamœba, Protogenes, Protomyxa, etc.) in my Monograph of the Monera,(15) and have briefly mentioned the most important facts in the eighth chapter (vol. i. pp. 183-187). Therefore, only by way of a specimen, I here repeat the drawing of the fresh-water Protamœba (Fig. 8). The history of the life of an orange-red Protomyxa adrantiaca, which I observed at Lanzerote, one of the Canary Islands, is given in Plate I. (see its explanation in the Appendix). Besides this, I here add a drawing of the form of Bathybius, that remarkable Moneron discovered by Huxley, which lives in the greatest depths of the sea in the shape of naked lumps of protoplasm and reticular mucus (vol. i. p. 344).

The Amœbæ of the present day, and the organisms most closely connected with them, Arcellidæ and Gregarinæ, which we here unite as a second class of Protista under the name of Amœboidea (Protoplasta), present no fewer genealogical difficulties than the Monera. These primary creatures are at present usually placed in the animal kingdom without its in reality being understood why. For simple naked cells—that is, shell-less plastids with a kernel—occur as well among real plants as real animals. The generative cells, for example, in many Algæ (spores and eggs) exist for a longer or shorter time in water in the form of naked cells with a kernel, which cannot be distinguished at all from the naked eggs of many animals (for example, those of the Siphonophorous Medusæ). (Compare the figure of a naked egg of a bladder-wrack in Chapter xvii. p. 90.) In reality every naked simple cell, whether it proceeds from an animal or vegetable body, cannot be distinguished from an independent Amœba. For an Amœba is nothing but a simple primary cell, a naked little lump of cell-matter, or plasma, containing a kernel. The contractility of this plasma, which the free Amœba shows in stretching out and drawing in its changing processes, is a general vital property of the organic plasma of all animal as well as of all vegetable plastids. When a freely moving Amœba, which perpetually changes its form, passes into a state of rest, it draws itself together into the form of a globule, and surrounds itself with a secreted membrane. It can then be as little distinguished from an animal egg as from a simple globular vegetable cell

Naked cells, with kernels, like those represented in Fig. 10 B, which are continuously changing, stretching out and drawing in formless, finger-like processes, and which are on this account called amœboid, are found frequently and widely dispersed in fresh water and in the sea; nay, are even found creeping on land. They take their food in the same way as was previously described in the case of the Protamœba (vol. i. p. 186). Their propagation by division can sometimes be observed. (Fig. 10 C, D.) I have described the processes in an earlier chapter (vol. i. p. 187). Many of these formless Amœbæ have lately been recognized as the early stages of development of other Protista (especially the Myxomycetæ), or as the freed cells of lower animals and plants. The colourless blood-cells of animals, for example, those of human blood, cannot be distinguished from Amœbæ. They, like the latter, can receive solid corpuscles into their interior, as I was the first to show by feeding them with finely divided colouring matters (Gen. Morph. i. 271). However, other Amœbæ (like the one given in Fig. 10) seem to be independent "good species," since they propagate themselves unchanged throughout many generations. Besides the real, or naked, Amœbæ (Gymnamœbæ), we also find widely diffused in fresh water case-bearing Amœbæ (Lepamœbæ), whose naked plasma body is partially protected by a more or less solid shell (Arcella), sometimes even by a case (Difflugia) composed of small stones. Lastly, we frequently find in the body of many lower animals parasitic Amœbæ (Gregarinæ), which, adapting themselves to a parasitic life, have surrounded their plasma-body with a delicate closed membrane.

The simple naked Amœbæ are, next to the Monera, the most important of all organisms to the whole science of biology, and especially to general genealogy. For it is evident that the Amœbæ originally arose out of simple Monera (Protamœbæ), by the important process of segregation taking place in their homogeneous viscid body—the differentiation of an inner kernel from the surrounding plasma. By this means the great progress from a simple cytod (without kernel) into a real cell (with kernel) was accomplished (compare Fig. 8 A and Fig. 10 B). As some of these cells at an early stage encased themselves by secreting a hardened membrane, they formed the first vegetable cells, while others, remaining naked, developed into the first aggregates of animal cells. The presence or absence of an encircling hard membrane forms the most important, although by no means the entire, difference of form between animal and vegetable cells. As vegetable cells even at an early stage enclose themselves within their hard, thick, and solid cellular shell, like that of the Amœbæ in a state of rest (Fig. 10 A), they remain more independent and less accessible to the influences of the outer world than are the soft animal cells, which are in most cases naked, or merely covered by a thin pliable membrane. But in consequence of this the vegetable cells cannot combine, as do the animal cells, for the construction of higher and composite fibrous tracts, for example, the nervous and muscular tissues. It is probable that, in the case of the most ancient single-celled organisms, there must have developed at an early stage the very important difference in the animal and vegetable mode of receiving food. The most ancient single-celled animals, being naked cells, could admit solid particles into the interior of their soft bodies, as do the Amœbæ (Fig. 10 B) and the colourless blood-cells; whereas the most ancient single-celled plants encased by their membranes were no longer able to do this, and could admit through it only fluid nutrition (by means of diffusion).

The Whip-swimmers (Flagellata), which we consider as a third class of the kingdom Protista, are of no less doubtful nature than the Amœbæ. They often show as close and important relations to the vegetable as to the animal kingdom. Some Flagellata at an early stage, when freely moving about, cannot be distinguished from real plants, especially from the spores of many Algæ; whereas others are directly allied to real animals, namely, to the fringed Infusoria (Ciliata). The Flagellata are simple cells which live in fresh or salt water, either singly or united in colonies. The characteristic part of their body is a very movable simple or compound whip-like appendage (whip, or flagellum) by means of which they actively swim about in the water. This class is divided into two orders. Among the fringed whip-swimmers (Cilioflagellata) there exists, in addition to the long whip, a short fringe of vibrating hairs, which is wanting in the unfringed whip-swimmers (Nudoflagellata). To the former belong the flint-shelled yellow Peridinia, which are largely active in causing the phosphorescence of the sea; to the latter belong the green Euglenæ, immense masses of which frequently make our ponds in spring quite green.

A very remarkable new form of Protista, which I have named Flimmer-ball (Magosphæra), I discovered only three years ago (in September, 1869), on the Norwegian coast (Fig. 12), and have more accurately described in my Biological Studies(15) (p. 137, Plate V.). Off the island of Gis-oe, near Bergen, I found swimming about, on the surface of the sea, extremely neat little balls composed of a number (between thirty and forty) of fringed pear-shaped cells, the pointed ends of which were united in the centre like radii. After a time the ball dissolved. The individual cells swarmed about independently in the water like fringed Infusoria, or Ciliata. These afterwards sank to the bottom, drew their fringes into their bodies, and gradually changed into the form of creeping Amœbæ (like Fig 10 B). These last afterwards encased themselves (as in Fig. 10 A), and then divided by repeated halvings into a large number of cells (exactly as in the case of the cleavage of the egg, Fig. 6, vol. i. p. 299). The cells became covered with vibratile hairs, broke through the case enclosing them, and now again swam about in the shape of a fringed ball (Fig. 12). This wonderful organism, which sometimes appears like a simple Amœba, sometimes as a single fringed cell, sometimes as a many-celled fringed ball, can evidently be classed with none of the other Protista, and must be considered as the representative of a new independent group. As this group stands midway between several Protista, and links them together, it may bear the name of Mediator, or Catallacta.

The Protista of the fifth class, the Tram-weavers, or Labyrinthuleæ, are of a no less puzzling nature; they were lately discovered by Cienkowski on piles in sea water (Fig. 13). They are spindle-shaped cells, mostly of a yellow-ochre colour, which are sometimes united into a dense mass, sometimes move about in a very peculiar way. They form, in a manner not yet explained, a retiform frame of entangled threads (compared to a labyrinth), and on the dense filamentous "tramways" of this frame they glide about. From the shape of the cells of the Labyrinthuleæ we might consider them as the simplest plants, from their motion as the simplest animals, but in reality they are neither animals nor plants.

The Flint-cells (Diatomeæ), a sixth class of Protista, are perhaps the most closely related to the Labyrinthuleæ. These primary creatures—which at present are generally considered as plants, although some celebrated naturalists still look upon them as animals—inhabit the sea and fresh waters in immense masses, and offer an endless variety of the most elegant forms. They are mostly small microscopic cells, which either live singly (Fig. 14), or united in great numbers, and occur either attached to objects, or glide and creep about in a peculiar manner. Their soft cell-substance, which is of a characteristic brownish yellow colour, is always enclosed by a solid and hard flinty shell, possessing the neatest and most varied forms. This flinty covering is open to the exterior only by one or two slits, through which the enclosed soft plasma-body communicates with the outer world. The flinty cases are found petrified in masses, and many rocks—for example, the Tripoli slate polish, the Swedish mountain meal, etc.,—are in a great measure composed of them.

A seventh class of Protista is formed by the remarkable Slime-moulds (Myxomycetes). They were formerly universally considered as plants, as real Fungi, until ten years ago the botanist De Bary, by discovering their ontogeny, proved them to be quite distinct from Fungi, and rather to be akin to the lower animals. The mature body is a roundish bladder, often several inches in size, filled with fine spore-dust and soft flakes (Fig. 15), as in the case of the well-known puff-balls (Gastromycetes). However, the characteristic cellular threads, or hyphæ, of a real fungus do not arise from the germinal corpuscles, or spores, of the Myxomycetes, but merely naked masses of plasma, or cells, which at first swim about in the form of Flagellata (Fig. 11), afterwards creep about like the Amœbæ (Fig. 10 B), and finally combine with others of the same kind to form large masses of "slime," or "plasmodia." Out of these, again, there arises, by-and-by, the bladder-shaped fruit-body. Many of my readers probably know one of these plasmodia, the Æthalium septicum, which in summer forms a beautiful yellow mass of soft mucus, often several feet in breadth, known by the name of "tan flowers," and penetrates tan-heaps and tan-beds. At an early stage these slimy, freely-creeping Myxomycetes, which live for the most part in damp forests, upon decaying vegetable substances, bark of trees, etc., are with equal justice or injustice declared by zoologists to be animals, while in the mature, bladder-shaped condition of fructification they are by botanists defined as plants.

The nature of the Ray-streamers (Rhizopoda), the eighth class of the kingdom Protista, is equally obscure. These remarkable organisms have peopled the sea from the most ancient times of the organic history of the earth, in an immense variety of forms, sometimes creeping at the bottom of the sea, sometimes swimming on the surface. Only very few live in fresh water (Gromia, Actinosphærium). Most of them possess solid calcareous or flinty shells of an extremely beautiful construction, which can be perfectly preserved in a fossil state. They have frequently accumulated in such huge numbers as to form mountain masses, although the single individuals are very small, and often scarcely visible, or completely invisible to the naked eye. A very few attain the diameter of a few lines, or even as much as a couple of inches. The name which the class bears is given because thousands of exceedingly fine threads of protoplasm radiate from the entire surface of their naked slimy body; these rays are quasi-feet, or pseudopodia, which branch off like roots (whence the term Rhizopoda, signifying root-footed), unite like nets, and are observed continually to change form, as in the case of the simpler plasmic feet of the Amœboidea, or Protoplasts. These ever-changing little pseudo-feet serve both for locomotion and for taking food.

The class of the Rhizopoda is divided into three different legions, viz. the chamber-shells, or Acyttaria, the sun-animalcules, or Heliozoa, and the basket-shells, or Radiolaria. The Chamber-shells (Acyttaria) constitute the first and lowest of these three legions; for the whole of their soft body consists merely of simple mucous or slimy cell-matter, or protoplasm, which has not differentiated into cells. However, in spite of this most primitive nature of body, most of the Acyttaria secrete a solid shell composed of calcareous earth, which presents a great variety of exquisite forms. In the more ancient and more simple Acyttaria this shell is a simple chamber, bell-shaped, tubular, or like the shell of a snail, from the mouth of which a bundle of plasmic threads issues. In contrast to these single-chambered forms (Monothalamia), the many-chambered forms (Polythalamia)—to which the great majority of the Acyttaria belong—possess a house, which is composed in an artistic manner of numerous chambers. These chambers sometimes lie in a row one behind the other, sometimes in concentric circles or spirals, in the form of a ring round a central point, and then frequently one above another in many tiers, like the boxes of an amphitheatre. This formation, for example, is found in the nummulites, whose calcareous shells, of the size of a lentil, have accumulated to the number of millions, and form whole mountains on the shores of the Mediterranean. The stones of which some of the Egyptian pyramids are built consist of such nummulitic limestone. In most cases the chambers of the shells of the Polythalamia are wound round one another in a spiral line. The chambers are connected with one another by passages and doors, like rooms of a large palace, and are generally open towards the outside by numerous little windows, out of which the plasmic body can stream or strain forth its little pseudo-feet, or rays of slime, which are always changing form. But in spite of the exceedingly complicated and elegant structure of this calcareous labyrinth, in spite of the endless variety in the structure and the decoration of its numerous chambers, and in spite of the regularity and elegance of their execution, the whole of this artistic palace is found to be the secreted product of a perfectly formless, slimy mass, devoid of any component parts! Verily, if the whole of the recent anatomy of animal and vegetable textures did not support our theory of plastids, if all its important results did not unanimously corroborate the fact that the whole miracle of vital phenomena and vital forms is traceable to the active agency of the formless albuminous combinations of protoplasm, the Polythalamia alone would secure the triumph of that theory. For we may here at any moment, by means of the microscope, point out the wonderful fact, first established by Dujardin and Max Schulze, that the formless mucus of the soft plasma-body, this true "matter of life," is able to secrete the neatest, most regular, and most complicated structures. This secretive skill is simply a result of inherited adaptation, and by it we learn to understand how this same "primæval slime"—this same protoplasm—can produce in the bodies of animals and plants the most different and most complicated cellular forms.

It is, moreover, a matter of special interest that the most ancient organism, the remains of which are found in a petrified condition, belongs to the Polythalamia. This organism is the "Canadian Life's-dawn" (Eozoon canadense), which has already been mentioned, and which was found a few years ago in the Ottawa formation (in the deepest strata of the Laurentian system), on the Ottawa river in Canada. If we expected to find organic remains at all in these most ancient deposits of the primordial period, we should certainly look for such of the most simple Protista as are covered with a solid shell, and in the organization of which the difference between animal and plant is as yet not indicated.

We know of but few species of the Sun-animalcules (Heliozoa), the second class of the Rhizopoda. One species is very frequently found in our fresh waters. It was observed even in the last century by a clergyman in Dantzig, Eichhorn by name, and it has been called after him, Actinosphærium Eichhornii. To the naked eye it appears as a gelatinous grey globule of mucus, about the size of a pin's head. Looking at it through the microscope, we see hundreds or thousands of fine mucous threads radiating from the central plasma body, and perceive that the inner layer of its cell-substance is different from the outer layer, which forms a bladder-like membrane. In consequence of its structure, this, the little sun-animalcule, although wanting a shell, really rises above the structureless Acyttaria, and forms the transition from these to the Radiolaria. The genus Cystophrys is of a nature akin to it.

The Basket-shells (Radiolaria) form the third and last class of the Rhizopoda. Their lower forms are closely allied to the Heliozoa and Acyttaria, whereas their higher forms rise far above them. They are essentially distinguished from both by the fact that the central part of their body is composed of many cells, and surrounded by a solid membrane. This closed "central capsule," generally of a globular shape, is covered by a mucous layer of plasma, out of which there radiate on all sides thousands of exceedingly fine threads, the branching and confluent so-called pseudopodia. Between these are scattered numerous yellow cells of unknown function, containing grains of starch. Most Radiolaria are characterized by a highly developed skeleton, which consists of flint, and displays a wonderful richness of the neatest and most curious forms. Sometimes this flinty skeleton forms a simple trellice-work ball (Fig. 16 s), sometimes a marvellous system of several concentric trelliced balls, encased in one another, and connected by radial staves. In most cases delicate spikes, which are frequently branched like a tree, radiate from the surface of the balls. In other cases the whole skeleton consists of only one flinty star, and is then generally composed of twenty staves, distributed according to definite mathematical laws, and united in a common central point. The skeletons of other Radiolaria again form symmetrical many-chambered structures, as in the case of the Polythalamia. Perhaps no other group of organisms develop in the formation of their skeletons such an amount of various fundamental forms, such geometrical regularity, and such elegant architecture. Most of the forms as yet discovered, I have given in the atlas accompanying my Monograph of the Radiolaria.(23) Here I shall only give as an example the picture of one of the simplest forms, the Cyrtidosphæra echinoides of Nice. The skeleton in this case consists only of a simple trelliced ball (s), with short radial spikes (a), which loosely surround the central capsule (c). Out of the mucous covering, enclosing the latter, radiate a great number of delicate little pseudopodia (p), which are partly drawn back underneath the shell, and fused into a lumpy mass of mucus. Between these are scattered a number of yellow cells (l).

Most Acyttaria live only at the bottom of the sea, on stones and seaweeds, or creep about in sand and mud by means of their pseudopodia, but most Radiolaria swim on the surface of the sea by means of long pseudopodia extending in all directions. They live together there in immense numbers, but are mostly so small that they have been almost completely overlooked, and have only become accurately known during the last fourteen years. Certain Radiolaria living in communities (Polycyttaria) form gelatinous lumps of some lines in diameter. On the other hand, most of those living isolated (Monocyttaria) are invisible to the naked eye; but still their petrified shells are found accumulated in such masses that in many places they form entire mountains; for example, the Nicobar Islands in the Indian Archipelago, and the Island of Barbadoes in the Antilles.

As most readers are probably but little acquainted with the eight classes of the Protista just mentioned, I shall now add some further general observations on their natural history. The great majority of all Protista live in the sea, some swimming freely on the surface, some creeping at the bottom, and others attached to stones, shells, plants, etc. Many species of Protista also live in fresh water, but only a very small number on dry land (for example, Myxomycetes and some Protoplasta). Most of them can be seen only through the microscope, except when millions of individuals are found accumulated. Only a few of them attain a diameter of some lines, or as much as an inch. What they lack in size of body they make up for by producing astonishing numbers of individuals, and they very considerably influence in this way the economy of nature. The imperishable remains of dead Protista, for instance, the flinty shells of the Diatomeæ and Radiolaria and the calcareous shells of the Acyttaria, often form large rock masses.

In regard to their vital phenomena, especially those of nutrition and propagation, some Protista are more allied to plants, others more to animals. Both in their mode of taking food and in the chemical changes of their living substance, they sometimes more resemble the lower animals, at others the lower plants. Free locomotion is possessed by many Protista, while others are without it; but this does not constitute a characteristic distinction, as we know of undoubted animals which entirely lack free locomotion, and of genuine plants which possess it. All Protista have a soul—that is to say, are "animate"—as well as all animals and all plants. The soul's activity in the Protista manifests itself in their irritability, that is, in the movements and other changes which take place in consequence of mechanical, electrical, and chemical irritation of their contractile protoplasm. Consciousness and the capability of will and thought are probably wanting in all Protista. However, the same qualities are in the same degree also wanting in many of the lower animals, whereas many of the higher animals in these respects are scarcely inferior to the lower races of human beings. In the Protista, as in all other organisms, the activities of the soul are traceable to molecular motions in the protoplasm.

The most important physiological characteristic of the kingdom Protista lies in the exclusively non-sexual propagation of all the organisms belonging to it. The higher animals and plants multiply almost exclusively in a sexual manner. The lower animals and plants multiply also, in many cases, in a non-sexual manner, by division, the formation of buds, the formation of germs, etc. But sexual propagation almost always exists by the side of it, and often regularly alternates with it in succeeding generations (Metagenesis, vol. i. p. 206). All Protista, on the other hand, propagate themselves exclusively in a non-sexual manner, and in fact, the distinction of the two sexes among them has not been effected—there are neither male nor female Protista.

The Protista in regard to their vital phenomena stand midway between animals and plants, that is to say, between their lowest forms; and the same must be said in regard to the chemical composition of their bodies. One of the most important distinctions between the chemical composition of animal and vegetable bodies consists in the characteristic formation of the skeleton. The skeleton, or the solid scaffolding of the body in most genuine plants, consists of a substance called cellulose, devoid of nitrogen, but secreted by the nitrogenous cell-substance, or protoplasm. In most genuine animals, on the other hand, the skeleton generally consists either of nitrogenous combinations (chitin, etc.) or of calcareous earth. In this respect some Protista are more like plants, others more like animals. In many of them the skeleton is principally or entirely formed of calcareous earth, which is met with both in animal and vegetable bodies. But the active vital substance in all cases is the mucous protoplasm.

In regard to the form of the Protista, it is to be remarked that the individuality of their body almost always remains at an extremely low stage of development. Very many Protista remain for life simple plastids or individuals of the first order. Others, indeed, form colonies or republics of plastids by the union of several individuals. But even these higher individuals of the second order, formed by the combination of simple plastids, for the most part remain at a very low stage of development. The members of such communities among the Protista remain very similar one to another, and never, or only in a slight degree, commence a division of labour, and are consequently as little able to render their community fit for higher functions as are, for example, the savages of Australia. The community of the plastids remains in most cases very loose, and each single plastid retains in a great measure its own individual independence.

A second structural characteristic, which next to their low stage of individuality especially distinguishes the Protista, is the low stage of development of their stereometrical fundamental forms. As I have shown in my theory of fundamental forms (in the fourth book of the General Morphology), a definite geometrical fundamental form can be pointed out in most organisms, both in the general form of the body and in the form of the individual parts. This ideal fundamental form, or type, which is determined by the number, position, combination, and differentiation of the component parts, stands in just the same relation to the real organic form as the ideal geometrical fundamental form of crystals does to their imperfect real form. In most bodies and parts of the bodies of animals and plants this fundamental form is a pyramid. It is a regular pyramid in the so-called "regular radiate" forms, and an irregular pyramid in the more highly differentiated, so-called "bilaterally symmetrical" forms. (Compare the plates in the first volume of my General Morphology, pp. 556-558.) Among the Protista this pyramidal type, which prevails in the animal and vegetable kingdom, is on the whole rare, and instead of it we have either quite irregular (amorphous) or more simple, regular geometrical types; especially frequent are the sphere, the cylinder, the ellipsoid, the spheroid, the double cone, the cone, the regular polygon (tetrahedron, hexahedron, octahedron, dodecahedron, icosahedron), etc. All the fundamental forms of the pro-morphological system, which are of a low rank in that system, prevail in the

Protista. However, in many Protista there occur also the higher, regular, and bilateral types, fundamental forms which predominate in the animal and vegetable kingdoms. In this respect some of the Protista are frequently more closely allied to animals (as the Acyttaria), others more so to plants (as the Radiolaria).

With regard to the palæontological development of the kingdom Protista, we may form various, but necessarily very unsafe, genealogical hypotheses. Perhaps the individual classes of the kingdom are independent tribes, or phyla, which have developed independently of one another and independently of the animal and the vegetable kingdoms. Even if we adopt the monophyletic hypothesis of descent, and maintain a common origin from a single form of Moneron for all organisms, without exception, which ever have lived and still live upon the earth, even in this case the connection of the neutral Protista on the one hand with the vegetable kingdom, and on the other hand with the animal kingdom, must be considered as very vague. We must regard them (compare p. 74) as lower offshoots which have developed directly out of the root of the great double-branched organic pedigree, or perhaps out of the lowest tribe of Protista, which may be supposed to have shot up midway between the two diverging high and vigorous trunks of the animal and vegetable kingdoms. The individual classes of the Protista, whether they are more closely connected at their roots in groups, or only form a loose bunch of root offsets, must in this case be regarded as having nothing to do either with the diverging groups of organisms belonging to the animal kingdom on the right, or to the vegetable kingdom on the left. They must be supposed to have retained the original simple character of the common primæval living thing more than have genuine animals and genuine plants.

But if we adopt the polyphyletic hypothesis of descent, we have to imagine a number of organic tribes, or phyla, which all shoot up by spontaneous generation out of the same ground, by the side of and independent of one another. (Compare p. 75.) In that case numbers of different Monera must have arisen by spontaneous generation whose differences would depend only upon slight, to us imperceptible, differences in their chemical composition, and consequently upon differences in their capability of development. A small number of Monera would then have given origin to the animal kingdom, and, again, a small number would have produced the vegetable kingdom. Between these two groups, however, there would have developed, independently of them, a large number of independent tribes, which have remained at a lower stage of organization, and which have neither developed into genuine plants nor into genuine animals.

A safe means of deciding between the monophyletic and polyphyletic hypotheses is as yet quite impossible, considering the imperfect state of our phylogenetic knowledge. The different groups of Protista, and those lowest forms of the animal kingdom and of the vegetable kingdom which are scarcely distinguishable from the Protista, show such a close connection with one another and such a confused mixture of characteristics, that at present any systematic division and arrangement of the groups of forms seem more or less artificial and forced. Hence the attempt here offered must be regarded as entirely provisional. But the more deeply we penetrate into the genealogical secrets of this obscure domain of inquiry, the more probable appears the idea that the vegetable kingdom and the animal kingdom are each of independent origin, and that midway between these two great pedigrees a number of other independent small groups of organisms have arisen by repeated acts of spontaneous generation, which on account of their indifferent neutral character, and in consequence of their mixture of animal and vegetable properties, may lay claim to the designation of independent Protista.

Thus, if we assume one entirely independent trunk for the vegetable kingdom, and a second for the animal kingdom, we may set up a number of independent stems of Protista, each of which has developed, quite independently of other stems and trunks, from a special archigonic form of Monera. In order to make this relation more clear, we may imagine the whole world of organisms as an immense meadow which is partially withered, and upon which two many-branched and mighty trees are standing, likewise partially withered. The two great trees represent the animal and vegetable kingdoms, their fresh and still green branches the living animals and plants; the dead branches with withered leaves represent the extinct groups. The withered grass of the meadow corresponds to the numerous extinct tribes, and the few stalks, still green, to the still living phyla of the kingdom Protista. But the common soil of the meadow, from which all have sprung up, is primæval by protoplasm.

PEDIGREE AND HISTORY OF THE VEGETABLE KINGDOM

Every attempt that we make to gain a knowledge of the pedigree of any small or large group of organisms related by blood must, in the first instance, start with the evidence afforded by the existing "natural system" of this group. For although the natural system of animals and plants will never become finally settled, but will always represent a merely approximate knowledge of true blood relationship, still it will always possess great importance as a hypothetical pedigree. It is true, by a "natural system" most zoologists and botanists only endeavour to express in a concise way the subjective conceptions which each has formed of the objective "form-relationships" of organisms. These form-relationships, however, as the reader has seen, are in reality the necessary result of true blood relationship. Consequently, every morphologist in promoting our knowledge of the natural system, at the same time promotes our knowledge of the pedigree, whether he wishes it or not. The more the natural system deserves its name, and the more firmly it is established upon the concordance of results obtained from the study of comparative anatomy, ontogeny, and palæontology, the more surely may we consider it as the approximate expression of the true pedigree of the organic world.

In entering upon the task contemplated in this chapter, the genealogy of the vegetable kingdom, we shall have, according to this principle, first to glance at the natural system of the vegetable kingdom as it is at present (with more or less important modifications) adopted by most botanists. According to the system generally in vogue, the whole series of vegetable forms is divided into two main groups. These main divisions, or sub-kingdoms, are the same as were distinguished more than a century ago by Charles Linnæus, the

47

founder of systematic natural history, and which he called Cryptogamia, or secretly-blossoming plants, and Phanerogamia, or openly-flowering plants. The latter, Linnæus, in his artificial system of plants, divided, according to the different number, formation, and combination of the anthers, and also according to the distribution of the sexual organs, into twenty-three different classes, and then added the Cryptogamia to these as the twenty-fourth and last class.

The Cryptogamia, the secretly-blossoming or flowerless plants, which were formerly but little observed, have in consequence of the careful investigations of recent times been proved to present such a great variety of forms, and such a marked difference in their coarser and finer structure, that we must distinguish no less than fourteen different classes of them; whereas the number of classes of flowering plants, or Phanerogamia, may be limited to four. However, these eighteen classes of the vegetable kingdom can again be naturally grouped in such a manner that we are able to distinguish in all six main divisions or branches of the vegetable kingdom. Two of these six branches belong to the flowering, and four to the flowerless plants. The table on page shows how the eighteen classes are distributed among the six branches, and how these again fall under the sub-kingdoms of the vegetable kingdom.

The one sub-kingdom of the Cryptogamia may now be naturally divided into two divisions, or sub-kingdoms, differing very essentially in their internal structure and in their external form, namely, the Thallus plants and the Prothallus plants. The group of Thallus plants comprises the two large branches of Tangles, or Algæ, which live in water, and the Thread-plants, or Inophytes (Lichens and Fungi), which grow on land, upon stones, bark of trees, upon decaying bodies, etc. The group of Prothallus plants, on the other hand, comprises the two branches of Mosses and Ferns, containing a great variety of forms.

All Thallus plants, or Thallophytes, can be directly recognized from the fact that the two morphological fundamental organs of all other plants, stem and leaves, cannot be distinguished in their structure. The complete body of all Algæ and of all Thread-plants is a mass composed of simple cells, which is called a lobe, or thallus. This thallus is as yet not differentiated into axial-organs (stem and root) and leaf-organs. On this account, as well as through many other peculiarities, the Thallophytes contrast strongly with all remaining plants—those comprised under the two sub-kingdoms of Prothallus plants and Flowering plants—and for this reason the two latter sub-kingdoms are frequently classed together under the name of Stemmed plants, or Cormophytes. The following table will explain the relation of these three sub-kingdoms to one another according to the two different views:—

I. Flowerless Plants.

(Cryptogamia)
A. Thallus Plants
(Thallophyta)
I. Thallus Plants
(Thallophyta)
B. Prothallus Plants
(Prothallophyta)
II. Stemmed Plants
(Cormophyta)
II. Flowering Plants
(Phanerogamia)
C. Flowering Plants
(Phanerogamia)

The stemmed plants, or Cormophytes, in the organization of which the difference of axial-organs (stem and root) and leaf-organs is already developed, form at present, and have, indeed, for a very long period formed, the principal portion of the vegetable world. However, this was not always the case. In fact, stemmed plants, not only of the flowering group, but even of the prothallus group, did not exist at all during that immeasurably long space of time which forms the beginning of the first great division of the organic history of the earth, under the name of the archilithic, or primordial period. The reader will recollect that during this period the Laurentian, Cambrian, and Silurian systems of strata were deposited, the thickness of which, taken as a whole, amounts to about 70,000 feet. Now, as the thickness of all the more recent superincumbent strata, from the Devonian to the deposits of the present time, taken together, amounts to only about 60,000 feet, we were enabled from this fact alone to draw the conclusion—which is probable also for other reasons— that the archilithic, or primordial, period was of longer duration than the whole succeeding period down to the present time. During the whole of this immeasurable space of time, which probably comprises many millions of centuries, vegetable life on our earth seems to have been represented exclusively by the sub-kingdom of Thallus plants, and, moreover, only by the class of marine Thallus plants, that is to say, the Algæ. At least all the petrified remains which are positively known to be of the primordial period belong exclusively to this class. As all the animal remains of this immense period also belong exclusively to animals that lived in water, we come to the conclusion that at that time organisms adapted to a life on land did not exist at all.

For these reasons the first and most imperfect of the great provinces or branches of the vegetable kingdom, the division of the Algæ, or Tangles, must be of special interest to us. But, in addition, there is the interest which this group offers when viewed by itself. In spite of the exceedingly simple composition of their constituent cells, which are but little differentiated, the Algæ show an extraordinary variety of different forms. To them belong the simplest and most imperfect of all forms, as well as very highly developed and peculiar forms. The different groups of Algæ are distinguished as much by size of body as by the perfection and variety of their outer form. At the lowest stage we find such species as the minute Protococcus, several hundred thousands of which occupy a space no larger than a pin's head. At the highest stage we marvel at the gigantic Macrocysts, which attain a length of from 300 to 400 feet, the longest of all forms in the vegetable kingdom. It is possible that a large portion of the coal has been formed out of Algæ. If not for these reasons, yet the Algæ must excite our special attention from the fact that they form the beginning of vegetable life, and contain the original forms of all other groups of plants, supposing that our monophyletic hypothesis of a common origin for all groups of plants is correct. (Compare p. 83.)

Most people living inland can form but a very imperfect idea of this exceedingly interesting branch of the vegetable kingdom, because they know only its proportionately small and simple representatives living in fresh water. The slimy green aquatic filaments and flakes of our pools and ditches and springs, the light green slimy coverings of all kinds of wood which have for any length of time been in contact with water, the yellowish green, frothy, and oozy growths of our village ponds, the green filaments resembling tufts of hair which occur everywhere in fresh water, stagnant and flowing, are for the most part composed of different species of Algæ. Only those who have visited the sea-shore, and wondered at the immense masses of cast-up seaweed, and who, from the rocky coast of the Mediterranean, have seen through the clear blue waters the beautifully-formed and highly-coloured vegetation of Algæ at the bottom, know how to estimate the importance of the class of Algæ. And yet, even these marine Algæ-forests of European shores, so rich in forms, give only a faint idea of the colossal forests of Sargasso in the Atlantic ocean, those immense banks of Algæ, covering a space of about 40,000 square miles—the same which made Columbus, on his voyage of discovery, believe that a continent was near. Similar but far more extensive forests of Algæ grew in the primæval ocean, probably in dense masses, and what countless generations of these archilithic Algæ have died out one after another is attested, among other facts, by the vast thickness of Silurian alum schists in Sweden, the peculiar composition of which proceeds from those masses of submarine Algæ. According to the recently expressed opinion of Frederick Mohr, a geologist of Bonn, even the greater part of our coal seams have arisen out of the accumulated dead bodies of the Algæ forests of the ocean.

Within the branch of the Algæ we distinguish four different classes, each of which is again divided into several orders and families. These again contain a large number of different genera and species. We designate these four classes as Primæval Algæ, or Archephyceæ, Green Algæ, or Chlorophyceæ, Brown Algæ, or Phæophyceæ, and Red Algæ, or Rhodophyceæ.

The first class of Algæ, the Primæval Algæ (Archephyceæ), might also be called primæval plants, because they contain the simplest and most imperfect of all plants, and, among them, those most ancient of all vegetable organisms out of which all other plants have originated. To them therefore belong those most ancient of all vegetable Monera which arose by spontaneous generation in the beginning of the Laurentian period. Further, we have to reckon among them all those vegetable forms of the simplest organization which first developed out of the Monera in the Laurentian period, and which possessed the form of a single plastid. At first the entire body of one of these small primary plants consisted only of a most simple cytod (a plastid without kernel), and afterwards attained the higher form of a simple cell, by the separation of a kernel in the plasma. (Compare above, vol. i. p. 345.) Even at the present day there exist various most simple forms of Algæ which have deviated but little from the original primary plants. Among them are the Algæ of the families Codiolaceæ, Protococcaceæ, Desmidiaceæ, Palmellaceæ, Hydrodictyeæ, and several others. The remarkable group of Phycochromaceæ (Chroococcaceæ and Oscillarineæ) might also be comprised among them, unless we prefer to consider them as an independent tribe of the kingdom Protista.

The monoplastic Protophyta—that is, those primary Algæ formed by a single plastid—are of the greatest interest, because the vegetable organism in this case completes its whole course of life as a perfectly simple "individual of the first order," either as a cytod without kernel, or as a cell containing a kernel.

Among the primary plants consisting of a single cytod are the exceedingly remarkable Siphoneæ, which are of considerable size, and strangely "mimic" the forms of higher plants. Many of the Siphoneæ attain a size of several feet, and resemble an elegant moss (Bryopsis), or in some cases a perfect flowering plant with stalks, roots, and leaves (Caulerpa) (Fig. 17). Yet the whole of this large body, externally so variously differentiated, consists internally of an entirely simple sack, possessing the negative characters of a simple cytod.

Caulerpa denticulata

Fig. 17.—Caulerpa denticulata, a monoplastic Siphonean of the natural size. The entire branching primary plant, which appears to consist of a creeping stalk with fibrous roots and indented leaves, is in reality only a single plastid, and moreover a cytod (without a kernel), not even attaining the grade of a cell with nucleus.

These curious Siphoneæ, Vaucheriæ, and Caulerpæ show us to how great a degree of elaboration a single cytod, although a most simple individual of the first order, can develop by continuous adaptation to the relations of the outer world. Even the single-celled primary plants—which are distinguished from the monocytods by possessing a kernel—develop into a great variety of exquisite forms by adaptation; this is the case especially with the beautiful Desmidiaceæ, of which a species of Euastrum.

It is very probable that similar primæval plants, the soft body of which, however, was not capable of being preserved in a fossil state, at one time peopled the Laurentian primæval sea in great masses and varieties, and in a great abundance of forms, without, however, going beyond the stage of individuality of a simple plastid.

The group of Green Tangles (Chlorophyceæ), or Green Algæ (Cloroalgæ), are the second class, and the most closely allied to the primæval group. Like the majority of the Archephyceæ, all the Chlorophyceæ are coloured green, and by the same colouring matter—the substance called leaf-green, or chlorophyll—which colours the leaves of all the higher plants.

To this class belong, besides a great number of low marine Algæ, most of the Algæ of fresh water, the common water hair-weeds, or Confervæ, the green slime-balls, or Glœosphæræ, the bright green water-lettuce, or Ulva, which resembles a very thin and long lettuce leaf, and also numerous small microscopic algæ, dense masses of which form a light green shiny covering to all sorts of objects lying in water—wood, stones, etc.

These forms, however, rise above the simple primary Algæ in the composition and differentiation of their body. As the green Algæ, like the primæval Algæ, mostly possess a very soft body, they are but rarely capable of being petrified. However, it can scarcely be doubted that this class of Algæ—which was the first to develop out of the preceding one—most extensively and variously peopled the fresh and salt waters of the earth in early times.

In the third class, that of the Brown Tangles (Phæophyceæ), or Black Algæ (Fucoideæ), the branch of the Algæ attains its highest stage of development, at least in regard to size and body. The characteristic colour of the Fucoid is more or less dark brown, sometimes tending more to an olive green or yellowish green, sometimes more to a brownish red or black colour.

Among these are the largest of all Algæ, which are at the same time the longest of all plants, namely, the colossal giant Algæ, amongst which the Macrocystis pyrifera, on the coast of California, attains a length of 400 feet. Also, among our indigenous Algæ, the largest forms belong to this group. Especially I may mention here the stately sugar-tangle (Laminaria), whose slimy, olive green thallus-body, resembling gigantic leaves of from 10 to 15 feet in length, and from a half to one foot in breadth, are thrown up in great masses on the coasts of the North and Baltic seas.

To this class belongs also the bladder-wrack (Fucus vesiculosus) common in our seas, whose fork-shaped, deeply-cut leaves are kept floating on the water by numerous air bladders (as is the case, too, with many other brown Algæ). The freely floating Sargasso Alga (Sargasso bacciferum), which forms the meadows or forests of the Sargasso Sea, also belongs to this class. Although each individual of these large alga-trees is composed of many millions of cells, yet at the beginning of its existence it consists, like all higher plants, of a single cell—a simple egg. This egg—for example, in the case of our common bladder-wrack—is a naked, uncovered cell, and as such is so like the naked egg-cells of lower marine animals—for example, those of the Medusæ—that they might easily be mistaken one for another (Fig. 19).

It was probably the Fucoideæ, or Brown Algæ, which during the primordial period, to a great extent constituted the characteristic alga-forests of that immense space of time. Their petrified remains, especially those of the Silurian period, which have been preserved, can, it is true, give us but a faint idea of them, because the material of these Algæ, like that of most others, is ill-suited for preservation in a fossil state. As has already been remarked, a large portion of coal is perhaps composed of them.

Less important is the fourth class of Algæ, that of the Rose-coloured Algæ (Rhodophyceæ), or Red Sea-weeds (Florideæ). This class, it is true, presents a great number of different forms; but most of them are of much smaller size than the Brown Algæ. Although they are inferior to the latter in perfection and differentiation, they far surpass them in some other respects. To them belong the most beautiful and elegant of all Algæ, which on account of the fine plumose division of their leaf-like bodies, and also on account of their pure and delicate red colour, are among the most charming of plants. The characteristic red colour sometimes appears as a deep purple, sometimes as a glowing scarlet, sometimes as a delicate rose tint, and may verge into violet and bluish purple, or on the other hand into brown and green tints of marvellous splendour. Whoever has visited one of our sea-coast watering places, must have admired the lovely forms of the Florideæ, which are frequently dried on white paper and offered for sale.

Most of the Red Algæ are so delicate, that they are quite incapable of being petrified; this is the case with the splendid Ptilotes, Plocamia, Delesseria, etc. However, there are individual forms, like the Chondria and Sphærococca, which possess a harder thallus, often almost as hard as cartilage, and of these fossil remains have been preserved—principally in the Silurian, Devonian, and Carboniferous strata, and later in the oolites. It is probable that this class also had an important share in the composition of the archilithic Algæ flora.

If we now again take into consideration the flora of the primordial period, which was exclusively formed by the group of Algæ, we can see that it is not improbable that its four subordinate classes had a share in the composition of those submarine forests of the primæval oceans, similar to that which the four types of vegetation—trees with trunks, flowering shrubs, grass, and tender leaf-ferns and mosses—at present take in the composition of our recent land forests.

We may suppose that the submarine tree forests of the primordial period were formed by the huge Brown Algæ, or Fucoideæ. The many-coloured flowers at the foot of these gigantic trees were represented by the gay Red Algæ, or Florideæ. The green grass between was formed by the hair-like bunches of Green Algæ, or Chloroalgæ. Finally, the tender foliage of ferns and mosses, which at present cover the ground of our forests, fill the crevices left by other plants, and even settle on the trunks of the trees, at that time probably had representatives in the moss and fern-like Siphoneæ, in the Caulerpa and Bryopsis, from among the class of the primary Algæ, Protophyta, or Archephyceæ.

With regard to the relationships of the different classes of Algæ to one another and to other plants, it is exceedingly probable that the Primary Algæ, or Archephyceæ, as already remarked, form the common root of the pedigree, not merely for the different classes of Algæ, but for the whole vegetable kingdom. On this account they may with justice be designated as primæval plants, or Protophyta.

Out of the naked vegetable Monera, in the beginning of the Laurentian period, enclosed cytods were probably the first to arise (vol. i. p. 345), by the naked, structureless, albuminous substance of the Monera becoming condensed in the form of a pellicle on the surface, or by secreting a membrane. At a later period, out of these enclosed cytods genuine vegetable cells probably arose, as a kernel or nucleus separated itself in the interior from the surrounding cell-substance or plasma.

The three classes of Green Algæ, Brown Algæ, and Red Algæ, are perhaps three distinct classes, which have arisen independently of one another out of the common radical group of Primæval Algæ, and then developed themselves further (each according to its kind), and have variously branched off into orders and families. The Brown and Red Algæ possess no close blood relationship to the other classes of the vegetable kingdom. These latter have most probably arisen out of the Primæval Algæ, either directly or by the intermediate step of the Green Algæ.

It is probable that Mosses (out of which, at a later time, Ferns developed) proceeded from a group of Green Algæ, and that Fungi and Lichens proceeded from a group of Primæval Algæ. The Phanerogamia developed at a much later period out of Ferns.

As a second class of the Vegetable Kingdom we have above mentioned the Thread-plants (Inophyta). We understood by this term the two closely related classes of Lichens and Fungi. It is possible that these Thallus plants have not arisen out of the Primæval Algæ, but out of one or more Monera, which, independently of the latter, arose by spontaneous generation. It appears conceivable that many of the lowest Fungi, as for example, many ferment-causing fungi (forms of Micrococcus, etc.), owe their origin to a number of different archigonic Monera (that is, Monera originating by spontaneous generation).

In any case the Thread-plants cannot be considered as the progenitors of any of the higher vegetable classes. Lichens, as well as fungi, are distinct from the higher plants in the composition of their soft bodies, consisting as it does of a dense felt-work of very long, variously interwoven, and peculiar threads or chains of cells—the so-called hyphæ, on which account we distinguish them as a province under the name Thread-plants. From their peculiar nature they could not leave any important fossil remains, and consequently we can form only a very vague guess at their palæontological development.

The first class of Thread-plants, the Fungi, exhibit a very close relationship to the lowest Algæ; the Algo-fungi, or Phycomycetes (the Saprolegniæ and Peronosporæ) in reality only differ from the bladder-wracks and Siphoneæ (the Vaucheria and Caulerpa) mentioned previously by the want of leaf-green, or chlorophyll. But, on the other hand, all genuine Fungi have so many peculiarities, and deviate so much from other plants, especially in their mode of taking food, that they might be considered as an entirely distinct province of the vegetable kingdom.

Other plants live mostly upon inorganic food, upon simple combinations which they render more complicated. They produce protoplasm by the combination of water, carbonic acid, and ammonia. They take in carbonic acid and give out oxygen. But the Fungi, like animals, live upon organic food, consisting of complicated combinations of carbon, which they receive from other organisms and assimilate. They inhale oxygen and give out carbonic acid like animals. They also never form leaf-green, or chlorophyll, which is so characteristic of most other plants. In like manner they never produce starch. Hence many eminent botanists have repeatedly proposed to remove the Fungi completely out of the vegetable kingdom, and to regard them as a special and third kingdom, between that of animals and plants. By this means our kingdom of Protista would be considerably increased. The Fungi in this case would, in the first place, be allied to the so-called "slime moulds," or Myxomycetes (which, however, never form any hyphæ). But as many Fungi propagate in a sexual manner, and as most botanists, according to the prevalent opinion, look upon Fungi as genuine plants, we shall here leave them in the vegetable kingdom, and connect them with lichens, to which they are at all events most nearly related.

The phyletic origin of Fungi will probably long remain obscure. The close relationship already hinted at between the Phycomycetes and Siphoneæ (especially between the Saprolegniæ and Vaucheriæ) suggests to us that they are derived from the latter. Fungi would then have to be considered as

Algæ, which by adaptation to a parasitical life have become very peculiarly transformed. Many facts, however, support the supposition that the lowest fungi have originated independently from archigonic Monera.

The second class of Inophyta, the Lichens (Lichenes), are very remarkable in relation to phylogeny; for the surprising discoveries of late years have taught us that every Lichen is really composed of two distinct plants—of a low form of Alga (Nostochaceæ, Chroococcaceæ), and of a parasitic form of Fungus (Ascomycetes), which lives as a parasite upon the former, and upon the nutritive substances prepared by it. The green cells, containing chlorophyll (gonidia), which are found in every lichen, belong to the Alga. But the colourless threads (hyphæ) which, densely interwoven, form the principal mass of the body of Lichens, belong to the parasitic Fungus. But in all cases the two forms of plants—Fungus and Alga—which are always considered as members of two quite distinct provinces of the vegetable kingdom, are so firmly united, and so thoroughly interwoven, that nearly every one looks upon a Lichen as a single organism.

Most Lichens form small, more or less formless or irregularly indented, crust-like coverings to stones, bark of trees, etc. Their colour varies through all possible tints, from the purest white to yellow, red, green, brown, and the deepest black.

Many lichens are important in the economy of nature from the fact that they can settle in the driest and most barren localities, especially on naked rocks upon which no other plant can live. The hard black lava, which covers many square miles of ground in volcanic regions, and which for centuries frequently presents the most determined opposition to the life of every kind of vegetation, is always first occupied by Lichens. It is the white or grey Lichens (Stereocaulon) which, in the most desolate and barren fields of lava, always begin to prepare the naked rocky ground for cultivation, and conquer it for subsequent higher vegetation. Their decaying bodies form the first mould in which mosses, ferns, and flowering plants can afterwards take firm root. Hardy Lichens are also less affected by the severity of climate than any other plants. Hence the naked rocks, even in the highest mountains—for the most part covered by eternal snow, on which no plant could thrive—are encrusted by the dry bodies of Lichens.

Leaving now the Fungi, Lichens, and Algæ, which are comprised under the name of Thallus plants, we enter upon the second sub-kingdom of the vegetable kingdom, that of the Prothallus plants (Prothallophyta), which by some botanists are called phyllogonic Cryptogamia (in contradistinction to the Thallus plants, or thallogonic Cryptogamia). This sub-kingdom comprises the two provinces of Mosses and Ferns.

Here we meet with (except in a few of the lowest forms) the separation of the vegetable body into two different fundamental organs, axial-organs (stem and root) and leaves (or lateral organs). In this the Prothallus plants resemble the Flowering plants, and hence the two groups have recently often been classed together as stemmed plants, or Cormophytes.

But, on the other hand, Mosses and Ferns resemble the Thallus plants, in the absence of the development of flowers and seeds, and even Linnæus classed them with these, as Cryptogamia, in contradistinction to the plants forming seeds; that is, flowering plants (Anthophyta or Phanerogamia).

Under the name of "Prothallus plants" we combine the closely-related Mosses and Ferns, because both exhibit a peculiar and characteristic "alternation of generation" in the course of their individual development. For every species exhibits two different generations, of which the one is usually called the Prothallium, or Fore-growth, the other is spoken of as the Cormus, or actual Stem of the moss or fern.

The first and original generation, the Fore-growth, or Prothallus, also called Protonema, still remains in that lower stage of elaboration manifested throughout life by all Thallus plants; that is to say, stem and leaf-organs have as yet not differentiated, and the entire cell-mass of the Fore-growth corresponds to a simple thallus. The second and more perfect generation of mosses and ferns—the Stem, or Cormus—develops a much more highly elaborate body, which has differentiated into stalk and leaf (as in the case of flowering plants), except in the lowest mosses, where this generation also remains in the lower stage of the thallus.

With the exception of these latter forms the first generation of Mosses and Ferns (the thallus-shaped Fore-growth) always produces a second generation with stem and leaves; the latter in its turn produces the thallus of the first generation, and so on. Thus, in this case, as in the ordinary cases of alternation of generation in animals, the first generation is like the third, fifth, etc., the second like the fourth, sixth, etc. (Compare vol. i. p. 206).

Of the two main classes of Prothallus plants, the Mosses in general are at a much lower stage of development than the Ferns, and their lowest forms (especially in an anatomical respect) form the transition from the Thallus plants through the Algæ to Ferns. The genealogical connection of Mosses and Ferns which is indicated by this fact can, however, be inferred only from the case of the most imperfect forms of the two classes; for the more perfect and higher groups of mosses and ferns do not stand in any close relation to one another, and develop in completely opposite directions. In any case Mosses have arisen directly out of Thallus plants, and probably out of Green Algæ.

Ferns, on the other hand, are probably derived from extinct unknown Mosses, which were very nearly related to the lowest liverworts of the present day. In the history of creation, Ferns are of greater importance than Mosses.

The branch of Mosses (Muscinæ, also called Musci, or Bryophyta) contains the lower and more imperfect plants of the group of Prothallophytes, which as yet do not possess vessels. Their bodies are mostly so tender and perishable that they are very ill-suited for being preserved in a recognizable state as fossils. Hence the fossil remains of all classes of Mosses are rare and insignificant. It is probable that Mosses developed in very early times out of the Thallus plants, or, to be more precise, out of the Green Algæ. It is probable that in the primordial period there existed aquatic forms of transition from the latter to Mosses, and in the primary period to those living on land. The Mosses of the present day—out of the gradually differentiating development of which comparative anatomy may draw some inferences as to their genealogy—are divided into two different classes, namely: (1) Liverworts; (2) Leafy Mosses.

The first and oldest class of Mosses, which is directly allied to the Green Algæ, or Confervæ, is formed by the Liverworts (Hepaticæ, or Thallobrya). The mosses belonging to them are, for the most part, small and insignificant in form, and are little known. Their lowest forms still possess, in both generations, a simple thallus like the Thallus plants; as for example, the Ricciæ and Marchantiaceæ. But the more highly developed liverworts, the Jungermanniaceæ and those akin to them, gradually commence to differentiate stem and leaf, and their most highly-developed forms are closely allied to leaf-mosses. By this transitional series the liverworts show their direct derivation from the Thallophytes, and more especially from the Green Algæ.

The Mosses, which are generally the only ones known to the uninitiated—and which, in fact, form the principal portion of the whole branch—belong to the second class, or Leafy Mosses (Musci frondosi, called Musci in a narrow sense, also Phyllobrya). Among them are most of those pretty little plants which, united in dense groups, form the bright glossy carpet of moss in our woods, or which, in company with liverworts and lichens, cover the bark of trees. As reservoirs, carefully storing up moisture, they are of the greatest importance in the economy of nature. Wherever man mercilessly cuts down and destroys forests, there, as a consequence, disappear the leafy mosses which covered the bark of the trees, or, protected by their shade, clothed the ground, and filled the spaces between the larger plants. Together with the leafy mosses disappear the useful reservoirs which stored up rain and dew for times of drought. Thus arises a disastrous dryness of the ground, which prevents the growth of any rich vegetation. In the greater part of Southern Europe—in Greece, Italy, Sicily, and Spain—mosses have been destroyed by the inconsiderate extirpation of forests, and the ground has thereby been robbed of its most useful stores of moisture; once flourishing and rich tracts of land have been changed into dry and barren wastes. Unfortunately in Germany, also, this rude barbarism is beginning to prevail more and more. It is probable that the small frondose mosses have played this exceedingly important part in nature for a very long time, possibly from the beginning of the primary period. But as their tender bodies are as little suited as those of all other mosses for being preserved in a fossil state, palæontology can give us no information about this.

We learn from the science of petrifactions much more than we do in the case of Mosses of the importance which the second branch of Prothallus plants—that is, Ferns—have had in the history of the vegetable world. Ferns, or more strictly speaking, the "plants of the fern tribe" (Filicineæ, or Pterideæ, also called Pteridophyta, or Vascular Cryptogams), formed during an extremely long period, namely, during the whole primary or palæolithic period, the principal portion of the vegetable world, so that we may without hesitation call it the era of Fern Forests. From the beginning of the Devonian period, in which organisms living on land appeared for the first time, namely, during the deposits of the Devonian, Carboniferous, and Permian strata, plants like Ferns predominated so much over all others, that we are justified in giving this name to that period. In the stratifications just mentioned, but above all, in the immense layers of coal of the Carboniferous or coal period, we find such numerous and occasionally well preserved remains of Ferns, that we can form a tolerable vivid picture of the very peculiar land flora of the palæolithic period. In the year 1855 the total number of the then known palæolithic species of plants amounted to about a thousand, and among these there were no less than 872 Ferns. Among the remaining 128 species were 77 Gymnosperms (pines and palm-ferns), 40 Thallus plants (mostly Algæ), and about 20 not accurately definable Cormophyta (stem-plants).

As already remarked, Ferns probably developed out of the lower liverworts in the beginning of the primary period. In their organization Ferns rise considerably above Mosses, and in their more highly developed forms even approach the flowering plants. In Mosses, as in Thallus plants, the entire body is composed of almost equi-formal cells, little if at all differentiated; but in the tissues of Ferns we find those peculiarly differentiated strings of cells which are called the vessels of plants, and which are universally met with in flowering plants. Hence Ferns are sometimes united as "vascular Cryptogams" with Phanerogams, and the group so formed is contrasted as that of the "vascular plants" with "cellular plants,"—that is, with "cellular cryptogams" (Mosses and Thallus plants). This very important process in the organization of plants—the formation of vessels—first occurred, therefore, in the Devonian period, consequently in the beginning of the second and smaller half of the organic history of the earth.

The branch of Ferns, or Filicinæ, is divided into five distinct classes: (1) Frondose Ferns, or Pteridæ; (2) Reed Ferns, or Calamariæ; (3) Aquatic Ferns, or Rhizocarpeæ; (4) Snakes Tongues, or Ophioglossæ; and (5) Scale Ferns, or Lepidophyta. By far the most important of these five classes, and also the richest in forms, were first the Frondose Ferns, and then the Scale-ferns, which formed the principal portion of the palæolithic forests. The Reed Ferns, on the other hand, had at that time already somewhat diminished in number; and of the Aquatic Ferns, we do not even know with certainty whether they then existed. It is difficult for us to form any idea of the very peculiar character of those gloomy palæolithic fern forests, in which the whole of the gay abundance of flowers of our present flora was entirely wanting, and which were not enlivened by any birds. Of the flowering plants there then existed only the two lowest classes, the pines and palm ferns, with naked seeds, whose simple and insignificant blossoms scarcely deserve the name of flowers.

The phylogeny of Ferns, and of the Gymnosperms which have developed out of them, has been made especially clear by the excellent investigations which Edward Strasburger published in 1872, on "The Coniferæ and Gnetaceæ," as also "On Azolla." This thoughtful naturalist and Charles Martins, of Montpellier, are among the few botanists who have thoroughly understood the fundamental value of the Theory of Descent, and the mechanical-causal connection between ontogeny and phylogeny. The majority of botanists do not even yet know the important difference between homology and analogy, between the morphological and physiological comparison of parts—which has long since been recognized in zoology—but Strasburger has employed this distinction and the principle of evolution in his "Comparative Anatomy of the Gymnosperms," in order to sketch the outlines of the blood relationship of this important group of plants.

The class among Ferns which has developed most directly out of the Liverworts is the class of real Ferns, in the narrow sense of the word, the Frondose Ferns (Filices, or Phyllopterides, also called Pteridæ). In the present flora of the temperate zones this class forms only a subordinate part, for it is in most cases represented only by low forms without trunks. But in the torrid zones, especially in the moist, steaming forests of tropical regions, this class presents us with the lofty palm-like fern trees. These beautiful tree-ferns of the present day, which form the chief ornament of our hot-houses, can however give us but a faint idea of the stately and splendid frondose ferns of the primary period, whose mighty trunks, densely crowded together, then formed entire forests. These trunks, accumulated in super-incumbent masses, are found in the coal seams of the Carboniferous period, and between them, in an excellent state of preservation, are found the impressions of the elegant fan-shaped leaves, crowning the top of the trunk in an umbrella-like bush. The varied outlines and the feather-like forms of these fronds, the elegant shape of the branching veins or bunches of vessels in their tender foliage, can still be as distinctly recognized in the impressions of the palæolithic fronds as in the fronds of ferns of the present day. In many cases even the clusters of fruit, which are distributed on the lower surface of the fronds, are distinctly preserved. After the Carboniferous period, the predominance of frondose ferns diminished, and towards the end of the secondary period they played almost as subordinate a part as they do at the present time.

The Calamariæ, Ophiglossæ, and Rhizocarpeæ seem to have developed as three diverging branches out of the Frondose Ferns, or Pteridæ. The Calamariæ, or Calamophyta, have remained at the lowest level among these three classes. The Calamariæ comprise three different orders, of which only one now exists, namely, the Horse-tails (Equisetaceæ). The two other orders, the Giant Reeds (Calamiteæ), and the Star-leaf Reeds (Asterophylliteæ), are long since extinct. All Calamariæ are characterized by a hollow and jointed stalk, stem, or trunk, upon which the branches and leaves (in cases where they exist) are set so as to encircle the jointed stem in whorls. The hollow joints of the stalk are separated from one another by partition walls. In Horse-tails and Calamiteæ the surface is traversed by longitudinal ribs running parallel, as in the case of a fluted column, and the outer skin contains so much silicious earth in the living forms, that it is used for cleansing and polishing. In the Asterophylliteæ, the star-shaped whorls of leaves were more strongly developed than in the two other orders. There exist, at present, of the Calamariæ only the insignificant Horse-tails (Equisetum), which grow in marshes and on moors; but during the whole of the primary and secondary periods they were represented by great trees of the genus Equisetites. There existed, at the same time, the closely related order of the Giant Reeds (Calamites), whose strong trunks grew to a height of about fifty feet. The order of the Asterophyllites, on the other hand, contained smaller and prettier plants, of a very peculiar form, and belongs exclusively to the primary period.

Among all Ferns, the history of the third class, that of the Root, or Aquatic Ferns (Rhizocarpeæ, or Hydropteridæ), is least known to us. In their structure these ferns, which live in fresh water, are on the one hand allied to the frond ferns, and on the other to the scaly ferns, but they are more closely related to the latter. Among them are the but little known moss ferns (Salvinia), clover ferns (Marsilea), and pill ferns (Pilularia) of our fresh waters; further, the large Azolla which floats in tropical ponds. Most of the aquatic ferns are of a delicate nature, and hence ill-suited for being petrified. This is probably the reason of their fossil remains being so scarce, and of the oldest of those known to us having been found in the Jura system. It is probable, however, that the class is much older, and that it was already developed during the palæolithic period out of other ferns by adaptation to an aquatic life.

The fourth class of ferns is formed by the Tongue Ferns (Ophiglossæ, or Glossopterides). These ferns, to which belongs the Botrychium, as well as the Ophioglossum (adder's-tongue) of our native genera, were formerly

61

considered as forming but a small subdivision of the frondose ferns. But they deserve to form a special class, because they represent important transitional forms from the Pterideæ and Lepidophytes towards higher plants, and must be regarded as among the direct progenitors of the flowering plants.

The fifth and last class is formed by the Scale Ferns (Lepidophytes, or Selagines). In the same way as the Ophioglossæ arose out of the frondose forms, the scale ferns arose out of the Ophioglossæ. They were more highly developed than all other ferns, and form the transition to flowering plants, which must have developed out of them. Next to the frondose ferns they took the largest part in the composition of the palæolithic fern forests. This class also contains, as does the class of reed ferns, three nearly related but still very different orders, of which only one now exists, the two others having become extinct towards the end of the Carboniferous period. The scaled ferns still existing belong to the order of the club-mosses (Lycopodiaceæ). They are mostly small, pretty moss-like plants, whose tender, many-branched stalk creeps in curves on the ground like a snake, and is densely encompassed and covered by small scaly leaves. The pretty creeping Lycopodium of our woods, which mountain tourists twine round their hats, is known to all, as also the still more delicate Selaginella, which under the name of creeping moss is used to adorn the soil of our hot-houses in the form of a thick carpet. The largest club-mosses of the present day are found in the Sunda Islands, where their stalks rise to the height of twenty-five feet, and attain half a foot in thickness. But in the primary and secondary periods even larger trees of this kind were widely distributed, the most ancient of which probably were the progenitors of the pines (Lycopodites). The most important dimensions were, however, attained by the class of scale trees (Lepidodendreæ), and by the seal trees (Sigillarieæ). These two orders, with a few species, appear in the Devonian period, but do not attain their immense and astonishing development until the Carboniferous period, and become extinct towards the end of it, or in the Permian period directly following upon it. The scale trees, or Lepidodendreæ, were probably more closely related to club-mosses than to Sigillarieæ. They grew into splendid, straight, unbranching trunks which divided at the top into numerous forked branches. They bore a large crown of scaly leaves, and like the trunk were marked in elegant spiral lines by the scars left at the base of the leaf stalks which had fallen off. We know of scale-marked trees from forty to sixty feet in length, and from twelve to fifteen feet in diameter at the root. Some trunks are said to be even more than a hundred feet in length. In the coal are found still larger accumulations of the no less highly developed but more slender trunks of the remarkable seal trees, Sigillarieæ, which in many places form the principal part of coal seams. Their roots were formerly described as quite a distinct vegetable form (under the name of Stigmaria). The Sigillarieæ are in many respects very like the scale-trees, but differ from them and from ferns in general in many ways. They were possibly closely related to the extinct Devonian Lycopterideæ, combining characteristic peculiarities of the club-mosses and the frondose ferns, which Strasburger considers as the hypothetical primary form of flowering plants.

In leaving the dense forests of the primary period, which were principally composed of frond ferns (Lepidodendreæ and Sigillarieæ), we pass onwards to the no less characteristic pine forests of the secondary period. Thus we leave the domain of the Cryptogamia, the plants forming neither flowers nor seeds, and enter the second main division of the vegetable kingdom, namely, the sub-kingdom of the Phanerogamia, flowering plants forming seeds. This division, so rich in forms, containing the principal portion of the present vegetable world, and especially the majority of plants living on land, is certainly of a much more recent date than the division of Cryptogamia. For it can have developed out of the latter only in the course of the palæolithic period. We can with full assurance maintain that, during the whole archilithic period, hence during the first and longer half of the organic history of the earth, no flowering plants as yet existed, and that they first developed during the primary period out of Cryptogamia of the fern kind. The anatomical and embryological relation of Phanerogamia to the latter is so close, that from it we can with certainty infer their genealogical connection, that is, their true blood relationship. Flowering plants cannot have directly arisen out of thallus plants, nor out of mosses; but only out of ferns, or Filicines. Most probably the scaled ferns, or Lepidophyta, and more especially amongst these the Lycopodiaceæ, forms closely related to the Selaginella of the present day, have been the direct progenitors of the Phanerogamia.

On account of its anatomical structure and its embryological development, the sub-kingdom of the Phanerogamia has for a long time been divided into two large branches, into the Gymnosperms, or plants with naked seeds, and the Angiosperms, or plants with enclosed seeds. The latter are in every respect more perfect and more highly organized than the former, and developed out of them only at a late date during the secondary period. The Gymnosperms, both anatomically and embryologically, form the transition group from Ferns to Angiosperms.

The lower, more imperfect, and the older of the two main classes of flowering plants, that of the Archispermeæ, or Gymnosperms (with naked seeds), attained its most varied development and widest distribution during the mesolithic or secondary epoch. It was no less characteristic of this period, than was the fern group of the preceding primary, and the Angiosperms of the succeeding tertiary, epoch. Hence we might call the secondary epoch that of Gymnosperms, or after its most important representatives, the era of Pine Forests. The Gymnosperms are divided into three classes: the Coniferæ, Cycadeæ, and Gnetaceæ. We find fossil remains of the pines, or Conifers, and of the Cycads, even in coal, and must infer from this that the transition from scaled ferns to Gymnosperms took place during the Coal, or possibly even in the Devonian period. However, the Gymnosperms play but a very subordinate part during the whole of the primary epoch, and do not predominate over Ferns until the beginning of the secondary epoch.

Of the two classes of Gymnosperms just mentioned, that of the Palm Ferns (Zamiæ, or Cycadeæ) stands at the lowest stage, and is directly allied to ferns, as the name implies, so that some botanists have actually included them in the fern group. In their external form they resemble palms, as well as tree ferns (or tree-like frond ferns), and are adorned by a crown of feathery leaves, which is placed either on a thick, short trunk, or on a slender, simple trunk like a pillar. At the present day this class, once so rich in forms, is but scantily represented by a few forms living in the torrid zones, namely, by the coniferous ferns (Zamia), the thick-trunked bread-tree (Encephalartos), and the slender-trunked Caffir bread-tree (Cycas). They may frequently be seen in hot-houses, and are generally mistaken for palms. A much greater variety of forms than occurs among the still existing palm ferns (Cycadeæ) is presented by the extinct and fossil Cycads, which occurred in great numbers more towards the middle of the secondary period, during the Jura, and which at that time principally determined the character of the forests.

The class of Pines, or coniferous trees (Coniferæ), has preserved down to our day a greater variety of forms than have the palm ferns. Even at the present time the trees belonging to it—cypresses, juniper trees, and trees of life (Thuja), the box and ginko trees (Salisburya), the araucaria and cedars, but above all the genus Pinus, which is so rich in forms, with its numerous and important species, spruces, pines, firs, larches, etc.—still play a very important part in the most different parts of the earth, and almost of themselves constitute extensive forests. Yet this development of pines seems but weak in comparison with the predominance which the class had attained over other plants during the early secondary period, that of the Trias. At that time mighty coniferous trees—with but proportionately few genera and species, but standing together in immense masses of individuals—formed the principal part of the mesolithic forests. This fact justifies us in calling the secondary period the "era of the pine forests," although the remains of Cycadeæ predominate over those of coniferous trees in the Jura period.2

From the pine forests of the mesolithic, or secondary period, we pass on into the leafy forests of the cænolithic, or tertiary period, and we arrive thus at the consideration of the sixth and last class of the vegetable kingdom, that of the Metaspermæ, Angiospermæ, or plants with enclosed seeds. The first certain and undoubted fossils of plants with enclosed seeds are found in the strata of the chalk system, and indeed we here find, side by side, remains of the two classes into which the main class of Angiosperms is generally divided, namely, the one seed-lobed plants, or monocotylæ, and the two seed-lobed plants, or dicotylæ. However, the whole group probably originated at an earlier period during the Trias. For we know of a number of doubtful and not accurately definable fossil remains of plants from the Oolitic and Trias (sic) periods, which some botanists consider to be Monocotylæ, whilst others consider them as Gymnosperms. In regard to the two classes of plants with enclosed seeds, the Monocotylæ and Dicotylæ, it is exceedingly probable that the Dicotyledons developed out of the Gnetaceæ, but that the Monocotyledons developed later out of a branch of the dicotyledons.

The class of one seed-lobed plants (Monocotylæ, or Monocotyledons, also called Endogenæ) comprises those flowering plants whose seeds possess but one germ leaf or seed lobe (cotyledon). Each whorl of its flower contains in most cases three leaves, and it is very probable that the mother plants of all Monocotyledons possessed a regular triple blossom. The leaves are mostly simple, and traversed by simple, straight bunches of vessels or "nerves." To this class belong the extensive families of the rushes, grasses, lilies, irids, and orchids, further a number of indigenous aquatic plants, the water-onions, sea grasses, etc., and finally the splendid and highly developed families of the Aroideæ and Pandaneæ, the bananas and palms. On the whole, the class of Monocotyledons—in spite of the great variety of forms which it developed, both in the tertiary and the present period—is much more simply organized than the class of the Dicotyledons, and its history of development also offers much less of interest. As their fossil remains are for the most part difficult to recognize, it still remains at present an open question in which of the three great secondary periods—the Trias, Jura, or chalk period—the Monocotyledons originated. At all events they existed in the chalk period as surely as did the Dicotyledons.

The second class of plants with enclosed seeds, the two seed-lobed (Dicotylæ, or Dicotyledons, also called Exogenæ) presents much greater historical and anatomical interest in the development of its subordinate groups. The flowering plants of this class generally possess, as their name indicates, two seed lobes or germ leaves (cotyledons). The number of leaves composing its blossom is generally not three, as in most Monocotyledons, but four, five, or a multiple of those numbers. Their leaves, moreover, are generally more highly differentiated and more composite than those of the Monocotyledons; they are traversed by crooked, branching bunches of vessels or "veins." To this class belong most of the leafed trees, and as they predominate in the tertiary period as well as, at present, over the

Gymnosperms and Ferns, we may call the cænolithic period that of leafed forests.

Although the majority of Dicotyledons belong to the most highly developed and most perfect plants, still the lowest division of them is directly allied to the Gymnosperms, and particularly to the Gnetaceæ. In the lower Dicotyledons, as in the case of the Monocotyledons, calyx and corolla are as yet not differentiated. Hence they are called Apetalous (Monochlamydeæ, or Apetalæ). This sub-class must therefore doubtless be looked upon as the original group of the Angiosperms, and existed probably even during the Trias and Jura periods. Among them are most of the leafed trees bearing catkins—birches and alders, willows and poplars, beeches and oaks; further, the plants of the nettle kind—nettles, hemp, and hops, figs, mulberries, and elms; finally, plants like the spurges, laurels, and amaranth.

It was not until the chalk period that the second and more perfect class of the Dicotyledons appeared, namely, the group with corollas (Dichlamydeæ, or Corollifloræ). These arose out of the Apetalæ from the simple cover of the blossoms of the latter becoming differentiated into calyx and corolla. The sub-class of the Corollifloræ is again divided into two large main divisions or legions, each of which contains a large number of different orders, families, genera, and species. The first legion bears the name of star-flowers, or Diapetalæ, the second that of the bell-flowers, or Gamopetalæ.

The lower and less perfect of the two legions of the Corollifloræ are the star-flowers (also called Diapetalæ or Dialypetalæ). To them belong the extensive families of the Umbelliferæ, or umbrella-worts (wild carrot, etc.), the Cruciferæ, or cruciform blossoms (cabbage, etc.); further, the Ranunculaceæ (buttercups) and Crassulaceæ, the Mallows and Geraniums, and, besides many others, the large group of Roses (which comprise, besides roses, most of our fruit trees), and the Pea-blossoms (containing, among others, beans, clover, genista, acacia, and mimosa). In all these Diapetalæ the blossom-leaves remain separate, and never grow together, as is the case in the Gamopetalæ. These latter developed first in the tertiary period out of the Diapetalæ, whereas the Diapetalæ appeared in the chalk period together with the Apetalæ.

The highest and most perfect group of the vegetable kingdom is formed by the second division of the Corollifloræ, namely, the legion of bell-flowers (Gamopetalæ, also called Monopetalæ or Sympetalæ). In this group the blossom-leaves, which in other plants generally remain separate, grow regularly together into a more or less bell-like, funnel-shaped, or tubular flower. To them belong, among others, the Bell-flowers and Convolvulus, Primroses and Heaths, Gentian and Honeysuckle, further the family of the Olives (olive trees, privet, elder, and ash), and finally, besides many other families, the extensive division of the Lip-blossoms (Labiatæ) and the

Composites. In these last the differentiation and perfection of the Phanerogamic blossoms attain their highest stage of development, and we must therefore place them at the head of the vegetable kingdom, as the most perfect of all plants. In accordance with this, the legion of the Gamopetalæ appear in the organic history of the earth later than all the main groups of the vegetable kingdom—in fact, not until the cænolithic or tertiary epoch. In the earliest tertiary period the legion is still very rare, but it gradually increases in the mid-tertiary, and attains its full development only in the latest tertiary and the quaternary period.

Now if, having reached our own time, we look back upon the whole history of the development of the vegetable kingdom, we cannot but perceive in it a grand confirmation of the Theory of Descent. The two great principles of organic development which have been pointed out as the necessary results of natural selection in the Struggle for Life, namely, the laws of differentiation and perfecting, manifest themselves everywhere in the development of the larger and smaller groups of the natural system of plants. In each larger or smaller period of the organic history of the earth, the vegetable kingdom increases both in variety and perfection, as a glance at Plate IV. will clearly show. During the whole of the long primordial period there existed only the lowest and most imperfect group, that of the Algæ. To these are added, in the primary period, the higher and more perfect Cryptogamia, especially the main-class of Ferns. During the coal period the Phanerogamia begin to develop out of the latter; at first, however, they are represented only by the lower main-class, that of Gymnosperms. It was not until the secondary period that the higher main-class, that of Angiosperms, arose out of them. Of these also there existed at first only the lower groups without distinct corollas, the Monocotyledons and the Apetalæ. It was not until the chalk period that the higher Corolliflore developed out of the latter. But even this most highly developed group is represented, in the chalk period, only by the lower stage of Star-flowers, or Diapetalæ, and only at quite a late date, in the tertiary period, did the more highly developed Bell-blossoms, Gamopetalæ, arise out of them, which at the same time are the most perfect of all flowering plants. Thus, in each succeeding later division of the organic history of the earth the vegetable kingdom gradually rose to a higher degree of perfection and variety.

PEDIGREE AND HISTORY OF THE ANIMAL KINGDOM

I. Animal-Plants and Worms.

The natural system of organisms which we must employ in the animal as well as in the vegetable kingdom, as a guide in our genealogical investigations, is in both cases of but recent origin, and essentially determined by the progress of comparative anatomy and ontogeny (the history of individual development) during the present century. Almost all the attempts at classification made in the last century followed the path of the artificial system, which was first established in a consistent manner by Charles Linnæus. The artificial system differs essentially from the natural one, in the fact that it does not make the whole organization and the internal structure (depending upon the blood relationship) the basis of classification, but only employs individual, and for the most part external, characteristics, which readily strike the eye. Thus Linnæus distinguished his twenty-four classes of the vegetable kingdom principally by the number, formation, and combination of the stamens. In like manner he distinguished six classes in the animal kingdom principally by the nature of the heart and blood. These six classes were: (1) Mammals; (2) Birds; (3) Amphibious Animals; (4) Fishes; (5) Insects; and (6) Worms.

But these six animal classes of Linnæus are by no means of equal value, and it was an important advance when, at the end of the last century, Lamarck comprised the first four classes as vertebrate animals (Vertebrata), and put them in contrast with the remaining animals (the insects and worms of Linnæus), of which he made a second main division—the invertebrate animals (Invertebrata). In reality Lamarck thus agreed with Aristotle, the father of Natural History, who had distinguished these two main groups, and called the former blood-bearing animals, the latter bloodless animals.

The next important progress towards a natural system of the animal kingdom was made some decades later by two most illustrious zoologists, Carl Ernst Bär and George Cuvier. As has already been remarked, they established, almost simultaneously and independently of one another, the proposition that it was necessary to distinguish several completely distinct main groups in the animal kingdom, each of which possessed an entirely peculiar type or structure (compare above, vol. i. p. 53). In each of these main divisions there is a tree-shaped and branching gradation from most simple and imperfect forms to those which are exceedingly composite and highly developed. The degree of development within each type is quite independent of the peculiar plan of structure, which forms the basis of the type and gives it a special characteristic. The "type" is determined by the peculiar relations in position of the most important parts of the body, and the manner in which the organs are connected. The degree of development, however, is dependent upon the greater or less division of labour among organs, and on the differentiation of the plastids and organs. This extremely important and fruitful idea was established by Bär, who relied more distinctly and thoroughly upon the history of individual development than did Cuvier. Cuvier based his argument upon the results of comparative anatomy. But neither of them recognized the true cause of the remarkable relationships pointed out by them, which is first revealed to us by the Theory of Descent. It shows us that the common type or plan of structure is determined by inheritance, and the degree of development or differentiation by adaptation. (Gen. Morph. ii. 10).

Both Bär and Cuvier distinguished four different types in the animal kingdom, and divided it accordingly into four great main divisions (branches or circles). The first of these is formed by the vertebrate animals (Vertebrata), and comprises Linnæus' first four classes—mammals, birds, amphibious animals, and fishes. The second type is formed by the articulated animals (Articulata), containing Linnæus' insects, consequently the six-legged insects, and also the myriopods, spiders, and crustacea, but besides these, a large number of the worms, especially the ringed worms. The third main division comprises the molluscous animals (Mollusca)—slugs, snails, mussels, and some kindred groups. Finally, the fourth and last circle of the animal kingdom comprises the various radiated animals (Radiata), which at first sight differ from the three preceding types by their radiated, flower-like form of body. For while the bodies of molluscs, articulated animals, and vertebrated animals consist of two symmetrical lateral halves—of two counterparts or antimera, of which the one is the mirror of the other—the bodies of the so-called radiated animals are composed of more than two, generally of four, five, or six counterparts grouped round a common central axis, as in the case of a flower. However striking this difference may seem at first, it is, in reality, a very subordinate one, and the radial form has by no means the same importance in all "radiated animals."

The establishment of these natural main groups or types of the animal kingdom by Bär and Cuvier was the greatest advance in the classification of animals since the time of Linnæus. The three groups of vertebrated animals, articulated animals, and molluscs are so much in accordance with nature that they are retained, even at the present day, little altered in extent. But a more accurate knowledge soon showed the utterly unnatural character of the group of the radiated animals. Leuckart, in 1848, first pointed out that two perfectly distinct types were confounded under the name, namely, the Star-fishes (Echinoderma)—the sea-stars, lily encrinites, sea-urchins, and sea-cucumbers; and, on the other hand, the Animal-plants, or Zoophytes (Cœlenterata or Zoophyta)—the sponges, corals, hood-jellies, and comb-jellies. At the same time, Siebold united the Infusoria with the Rhizopoda, under the name of Protozoa (lowest animals), into a special main division of the animal kingdom. By this the number of animal types was increased to six. It was finally increased to seven by the fact that modern zoologists separated the main division of the articulated animals into two groups: (a) those possessing articulated feet (Arthropoda), corresponding to Linnæus' Insects, namely, the Flies (with six legs), Myriopods, Spiders, and Crustacea; and (b) the footless Worms (Vermes), or those possessing non-articulated feet. These latter comprise only the real or genuine Worms (ring-worms, round worms, planarian worms, etc.), and therefore in no way correspond with the Worms of Linnæus, who had included the molluscs, the radiates, and many other lower animals under this name.

Thus, according to the views of modern zoologists, which are given in all recent manuals and treatises on zoology, the animal kingdom is composed of seven completely distinct main divisions or types, each of which is distinguished by a characteristic plan of structure peculiar to it, and perfectly distinct from every one of the others. In the natural system of the animal kingdom—which I shall now proceed to explain as its probable pedigree—I shall on the whole agree with this usual division, but not without some modifications, which I consider very important in connection with genealogy, and which are rendered absolutely necessary in consequence of our view as to the history of the development of animals.

We evidently obtain the greatest amount of information concerning the pedigree of the animal kingdom (as well as concerning that of the vegetable kingdom) from comparative anatomy and ontogeny. Besides these, palæontology also throws much valuable light upon the historical succession of many of the groups. From numerous facts in comparative anatomy, we may, in the first place, infer the common origin of all those animals which belong to one of the seven "types." For in spite of all the variety in the external form developed within each of these types, the essential relative position of the parts of the body which determines the type, is so constant, and agrees so completely in all the members of every type, that on account of their relations of form alone we are obliged to unite them, in the natural system, into a single main group. But we must certainly conclude, moreover, that this conjunction also has its expression in the pedigree of the animal kingdom. For the true cause of the intimate agreement in structure can only be the actual blood relationship. Hence we may, without further discussion, lay down the important proposition that all animals belonging to one and the same circle or type must be descended from one and the same original primary form. In other words, the idea of the circle or type, as it is employed in zoology since Bär and Cuvier's time to designate the few principal main groups or "sub-kingdoms" of the animal kingdoms, coincides with the idea of "tribe" or "phylum," as employed by the Theory of Descent.

If, then, we can trace all the varieties of animal forms to these seven fundamental forms, the following question next presents itself to us as a second phylogenetic problem—Where do these seven animal tribes come from? Are they seven original primary forms of an entirely independent origin, or are they also distantly related by blood to one another?

At first we might be inclined to answer this question in a polyphyletic sense, by saying that we must assume, for each of the seven great animal tribes, at least one independent primary form completely distinct from the others. On further considering this difficult problem, we arrive in the end at the notion of a monophyletic origin of the animal kingdom, viz., that these seven primary forms are connected at their lowest roots, and that they are derived from a single, common primæval form. In the animal as well as in the vegetable kingdom, when closely and accurately considered, the monophyletic hypothesis of descent is found to be more satisfactory than the polyphyletic hypothesis.

It is comparative ontogeny (embryology) which first and foremost leads to the assumption of the monophyletic origin of the whole animal kingdom (the Protista excepted of course). The zoologist who has thoughtfully compared the history of the individual development of various animals, and has understood the importance of the biogenetic principle (p. 33), cannot but be convinced that a common root must be assumed for the seven different animal tribes, and that all animals, including man, are derived from a single, common primary form. The result of the consideration of the facts of embryology, or ontogeny, is the following genealogical or phylogenetic hypothesis, which I have put forward and explained in detail in my "Philosophy of Calcareous Sponges" (Monograph of the Calcareous Sponges, vol. i. pp. 464, 465, etc.,—"the Theory of the Layers of the Embryo, and the Pedigree of Animals.")

The first stage of organic life in the Animal kingdom (as in the Vegetable and Protista kingdoms) was formed by perfectly simple Monera, originating by spontaneous generation. The former existence of this simplest animal form is, even at present, attested by the fact that the egg-cell of many animals loses its kernel directly after becoming fructified, and thus relapses to the lower stage of development of a cytod without a kernel, like a Moneron. This remarkable occurrence I have interpreted, according to the law of latent inheritance (vol. i. p. 205), as a phylogenetic relapse of the cellular form into the original form of a cytod. The Monerula, as we may call this egg-cytod without a kernel, repeats then, according to the biogenetic principle (vol. ii. p. 33), the most ancient of all animal forms, the common primary form of the animal kingdom, namely, the Moneron.

The second ontogenetic process consists in a new kernel being formed in the Monerula, or egg-cytod, which thus returns again to the value of a true egg-cell. According to this, we must look upon the simple animal cell, containing a kernel, or the single-celled primæval animal—which may still be seen in a living state in the Amœbæ of the present day—as the second step in the series of phylogenetic forms of the animal kingdom. Like the still living simple Amœbæ, and like the naked egg-cells of many lower animals (for example, of Sponges and Medusæ, etc.), which cannot be distinguished from them, the remote phyletic primary Amœbæ also were perfectly simple naked-cells, which moved about in the Laurentian primæval ocean, creeping by means of the ever-changing processes of their body-substance, and nourishing and propagating themselves in the same way as the Amœbæ of the present day. (Compare vol. i. p. 188, and vol. ii. p. 54.) The existence of this Amœba-like, single-celled primary form of the whole animal kingdom is unmistakably indicated by the exceedingly important fact that the egg of all animals, from those of sponges and worms up to those of the ant and man, is a simple cell.

Thirdly, from the "single-cell" state arose the simplest multicellular state, namely, a heap or a small community of simple, equiformal, and equivalent cells. Even at the present day, in the ontogenetic development of every animal egg-cell, there first arises a globular heap of equiformal naked cells, by the repeated self-division of the primary cell. (Compare vol. i. p. and the Frontispiece, Fig. 3.) We called this accumulation of cells the mulberry state (Morula), because it resembles a mulberry or blackberry. This Morula-body occurs in the same simple form in all the different tribes of animals, and on account of this most important circumstance we may infer—according to the biogenetic principle—that the most ancient, many-celled, primary form of the animal kingdom resembled a Morula like this, and was in fact a simple heap of Amœba-like primæval cells, one similar to the other. We shall call this most ancient community of Amœbæ—this most simple accumulation of animal cells—which is recapitulated in individual development by the Morula—the Synamœba.

Out of the Synamœbæ, in the early Laurentian period, there afterwards developed a fourth primary form of the animal kingdom, which we shall call the ciliated germ (Planæa). This arose out of the Synamœba by the outer cells on the surface of the cellular community beginning to extend vibrating fringes called cilia, and becoming "ciliated cells," and thus differentiating from the inner and unchanged cells. The Synamœbæ consisted of completely equi-formed and naked cells, and crept about slowly, at the bottom of the Laurentian primæval ocean, by means of movements like those of an Amœba. The Planæa, on the other hand, consisted of two kinds of different cells—inner ones like the Amœbæ, and external "ciliated cells." By the vibrating movements of the cilia the entire multicellular body acquired a more rapid and stronger motion, and passed over from the creeping to the swimming mode of locomotion. In exactly the same manner the Morula, in the ontogenesis of lower animals, still changes into a ciliated form of larva, which has been known, since the year 1847, under the name of Planula. This Planula is sometimes a globular, sometimes an oval body, which swims about in the water by means of a vibrating movement; the fringed (ciliated) and smaller cells of the surface differ from the larger inner cells, which are unfringed. (Fig. 4 of the Frontispiece.)

Out of this Planula, or fringed larva, there then develops, in animals of all tribes, an exceedingly important and interesting animal form, which, in my Monograph of the Calcareous Sponges, I have named Gastrula (that is, larva with a stomach or intestine). (Frontispiece, Fig. 5, 6). This Gastrula externally resembles the Planula, but differs essentially from it in the fact that it encloses a cavity which opens to the outside by a mouth. The cavity is the "primary intestine," or "primary stomach," the progaster, the first beginning of the alimentary canal; its opening is the "primary mouth" (prostoma). The wall of the progaster consists of two layers of cells: an outer layer of smaller ciliated cells (outer skin, or ectoderm), and of an inner layer of larger non-ciliated cells (inner skin, or entoderm). This exceedingly important larval form, the "Gastrula," makes its appearance in the ontogenesis of all tribes of animals—in Sponges, Medusæ, Corals, Worms, Sea-squirts, Radiated animals, Molluscs, and even in the lowest Vertebrata (Amphioxus: compare p. 200, Plate XII., Fig. B 4; see also in the same place the Ascidian, Fig. A 4).

From the ontogenetic occurrence of the Gastrula in the most different animal classes, from Zoophytes up to Vertebrata, we may, according to the biogenetic principle, safely draw the conclusion that during the Laurentian period there existed a common primary form of the six higher anima tribes, which in all essential points was formed like the Gastrula, and which we shall call the Gastræa. This Gastræa possessed a perfectly simple globular or oval body, which enclosed a simple cavity of like form, namely, the progaster; at one of the poles of the longitudinal axis the primary intestine opened by a mouth which served for the reception of nutrition. The body wall (which was also the intestinal wall) consisted of two layers of cells, the unfringed entoderm, or intestinal layer, and the fringed ectoderm, or skin-layer; by the motion of the cilia or fringes of the latter the Gastræa swam about freely in the Laurentian ocean. Even in those higher animals, in the ontogenesis of which the original Gastrula form has disappeared, according to the laws of abbreviated inheritance (vol. i. p. 212), the composition of the Gastræa body has been transmitted to the phase of development which directly arises out of the Morula. This phase is an oval or round disc consisting of two cell-layers or membranes: the outer cell-layer, the animal or dermal layer (ectoblast), corresponds to the ectoderm of the Gastræa; out of it develops the external, loose skin (epidermis), with its glands and appendages, as well as the central nervous system. The inner cell-layer, the vegetative or intestinal layer (hypoblast), is originally the entoderm of the Gastræa; out of it develops the inner membrane (epithelium) of the intestinal canal and its glands. (Compare my Monograph of the Calcareous Sponges, vol. i. p. 466, etc.)

By ontogeny we have already gained five primordial stages of development of the animal kingdom: (1) the Moneron; (2) the Amœba; (3) the Synamœba; (4) the Planæa; and (5) the Gastræa. The former existence of these five oldest primary forms, which succeeded one another, and which must have lived in the Laurentian period, follows as a consequence of the biogenetic principle; that is to say, from the parallelism and the mechanico-causal connection of ontogenesis and phylogenesis. (Compare vol. i. p. 309.) In our genealogical system of the animal kingdom we may class all these animal forms, long since extinct, and, which on account of the soft nature of their bodies could leave no fossil remains, among the tribe of Primæval animals (Protozoa), which also comprises the still living Infusoria and Gregarinæ.

The phyletic development of the six higher animal tribes, which are all derived from the Gastræa, deviated at this point in two directions. In other words, the Gastræads (as we may call the group of forms characterized by the Gastræa-type of structure), divided into two divergent lines or branches; the one branch of Gastræads gave up free locomotion, adhered to the bottom of the sea, and thus, by adopting an adhesive mode of life, gave rise to the Protascus, the common primary form of the Animal-plants (Zoophyta). The other branch of the Gastræads retained free locomotion, did not become adherent and later on developed into the Prothelmis, the common primary form of Worms (Vermes). (Compare p. 133.)

This latter tribe (as limited by modern zoology) is of the greatest interest in the study of genealogy. For among Worms, as we shall see later, there are, besides very numerous peculiar families, and besides many independent classes, also very remarkable forms, which may be considered as forms of direct transition to the four higher animal tribes. Both comparative anatomy and the ontogeny of these worms enable us to recognize in them the nearest blood relations of those extinct animal forms which were the original primary forms of the four higher animal tribes. Hence these latter, the Molluscs, Star-fishes, Articulated animals, and Vertebrate animals, do not stand in any close blood relationship to one another, but have originated independently in four different places out of the tribe of Worms. In this way comparative anatomy and phylogeny lead us to the monophyletic pedigree of the animal kingdom, the outlines of which are given on p. 133. According to it the seven phyla, or tribes, of the animal kingdom are of different value in regard to genealogy. The original primary group of the whole animal kingdom is formed by the Primæval animals (Protozoa), including the Infusoria and Gastræads. Out of these latter arose the two tribes of Animal-plants (Zoophyta) and Worms as diverging branches. Out of four different groups of the Worm tribe, the four higher tribes of the animal kingdom were developed—the Star-fishes (Echinoderma) and Insects (Arthropoda) on the one hand, and the Molluscs (Mollusca) and Vertebrated animals (Vertebrata) on the other.

Having thus sketched out the monophyletic pedigree of the animal kingdom in its most important features, we must now turn to a closer examination of the historical course of development which the seven tribes of the animal kingdom, and the classes distinguished in them, have passed through (p. 132). There is a much larger number of classes in the animal than in the vegetable kingdom, owing to the simple reason that the animal body, in consequence of its more varied and perfect vital activity, could differentiate and develop in very many more different directions than could the vegetable body. Thus, while we were able to divide the whole vegetable kingdom into six main classes and nineteen classes, we have to distinguish, at least, sixteen main classes and thirty-eight classes in the animal kingdom These are distributed among the seven different tribes of the animal kingdom in the way shown in the Systematic Survey.

The group of Primæval animals (Protozoa) within the compass which we here assign to this tribe, comprises the most ancient and the simplest primary forms of the animal kingdom; for example, the five oldest phyletic stages of development previously mentioned, and besides these the Infusoria and Gregarinæ, as well as all those imperfect animal forms, for which, on account of their simple and indifferent organization, no place can be found in any of the other six animal tribes. Most zoologists, in addition to these, Include among the Protozoa a larger or smaller portion of those lowest organisms, which we mentioned in our neutral kingdom of Protista (in Chapter XVI.). But these Protista, especially the large division of the Rhizopoda, which are so rich in forms, cannot be considered as real animals for reasons previously given. Hence, if we here leave them out of the question, we may accept two main classes or provinces of real Protozoa, namely, Egg animals (Ovularia) and Germ animals (Blastularia). To the former belong the three classes of Archezoa, Gregarinæ, and Infusoria, to the latter the two classes of Planæads and Gastræads.

The first province of the Protozoa consists of the Egg animals (Ovularia); we include among them all single-celled animals, all animals whose body, in the fully developed state, possesses the form-value of a simple plastid (of a cytod or a cell), also those simple animal forms whose body consists of an aggregation of several cells perfectly similar one to another.

The Archaic animals (Archezoa) form the first class in the series of Egg animals. It contains only the most simple and most ancient primary forms of the animal kingdom, whose former existence we have proved by means of the fundamental law of biogenesis; they are, (1) Animal Monera; (2) Animal Amœbæ; (3) Animal Synamœbæ. We may, if we choose, include among them a portion of the still living Monera and Amœbæ, but another portion (according to the discussion in Chapter XVI.) must on account of their neutral nature be considered as Protista, and a third portion, on account of their vegetable nature, must be considered as plants.

A second class of the egg animals consists of the Gregarines (Gregarinæ), which live as parasites in the intestines and body-cavities of many animals. Some of these Gregarines are perfectly simple cells like the Amœbæ; some form chains of two or three identical cells, one lying behind the other. They differ from the naked Amœbæ by possessing a thick, simple membrane, which surrounds their cell-body; they can be considered as animal Amœbæ which have adopted a parasitical mode of life, and in consequence have surrounded themselves with a secreted covering.

As a third class of egg animals, we adopt the real Infusoria (Infusoria), embracing those forms to which modern zoology almost universally limits this class of animals. The principal portion of them consists of the small ciliated Infusoria (Ciliata), which inhabit all the fresh and salt waters of the earth in great numbers, and which swim about by means of a delicate garb of vibratile fringes. A second and smaller division consists of the adherent sucking Infusoria (Acinetæ), which take their food by means of fine sucking-tubes. Although during the last thirty years numerous and very careful investigations have been made on these small animalcules,—which are mostly invisible to the naked eye,—still we are even now not very sure about their development and form-value. We do not even yet know whether the Infusoria are single or many-celled; but as no investigator has as yet proved their body to be a combination of cells, we are, in the mean time, justified in considering them as single-celled, like the Gregarines and the Amœbæ.

The second main class of primæval animals consists of the Germ animals

(Blastularia). This name we give to those extinct Protozoa which correspond to the two ontogenetic embryonic forms of the six higher animal tribes, namely, the Planula and the Gastrula. The body of these Blastularia, in a perfectly developed state, was composed of many cells, and these cells moreover differentiated—in two ways at least—into an external (animal or dermal) and an internal (vegetative or gastral) mass. Whether there still exist representatives of this group is uncertain. Their former existence is undoubtedly proved by the two exceedingly important ontogenetic animal forms which we have already described as Planula and Gastrula, and which still occur as a transient stage of development in the ontogeny of the most different tribes of animals. Corresponding to these, we may, according to the biogenetic principle, assume the former existence of two distinct classes of Blastularia, namely, the Planæada and Gastræada. The type of the Planæada is the Planæa—long since extinct—but whose historical portrait is still presented to us at the present day in the widely distributed ciliated larva (Planula). (Frontispiece, Fig. 4.) The type of the Gastræada is the Gastræa, of whose original nature the mouth-and-stomach larva (Gastrula), which recurs in the most different animal tribes, still gives a faithful representation. (Frontispiece Fig. 5, 6.) Out of the Gastræa, as we have previously mentioned, there were at one time developed two different primary forms, the Protascus and Prothelmis; the former must be looked upon as the primary form of the Zoophytes, the latter as the primary form of Worms. (Compare the enunciation of this hypothesis in my Monograph of the Calcareous Sponges, vol i. p. 464.)

The Animal-plants (Zoophyta, or Cœlenterata) which constitute the second tribe of the animal kingdom, rise considerably above the primitive animals in the characters of their whole organisation, while they remain far below most of the higher animals. For in the latter (with the exception only of the lowest forms) the four distinct functions of nutrition—namely, digestion, circulation of the blood, respiration, and excretion—are universally accomplished by four perfectly different systems of organs; by the intestines, the vascular system, the organs of respiration, and the urinary apparatus. In Zoophytes, however, these functions and their organs are not yet separate, and are all performed by a single system of alimentary canals, by the so-called gastro-vascular system, or the cœlenteric apparatus of the intestinal cavity. The mouth, which is also the anus, leads into a stomach, into which the other cavities of the body also open. In Zoophytes the body-cavity, or "cœloma," possessed by the four higher tribes of animals is still completely wanting, likewise the vascular system and blood, as also the organs of respiration, etc.

All Zoophytes live in water; most of them in the sea, only a very few in fresh water, such as fresh-water sponges (Spongilla) and some primæval polyps (Hydra, Cordylophora). A specimen of the pretty flower-like forms which are met with in great variety among Zoophytes is given on Plate VII. (Compare its explanation in the Appendix.)

The tribe of animal-plants, or Zoophytes, is divided into two distinct provinces, the Sponges, or Spongiæ, and the Sea-nettles, or Acalephæ (p. 144). The latter are much richer in forms and more highly organized than the former. In all Sponges the entire body, as well as the individual organs, are differentiated and perfected to a much less extent than in Sea-nettles. All Sponges lack the characteristic nettle-organs which all Sea-nettles possess.

The common primary form of all Zoophytes must be looked for in the Protascus, an animal form long since extinct, but whose existence is proved according to the biogenetic principle by the Ascula. This Ascula is an ontogenetical development form which, in Sponges as wellas in Sea-nettles, proceeds from the Gastrula. (Compare the Ascula of the calcareous sponge on the Frontispiece, Fig. 7, 8.) For after the Gastrula of zoophytes has for a time swum about in the water it sinks to the bottom, and there adheres by that pole of its axis which is opposite to the opening of the mouth. The external cells of the ectoderm draw in their vibrating, ciliary hairs, whereas, on the contrary, the inner cells of the entoderm begin to form them. Thus the Ascula, as we call this changed form of larva, is a simple sack, its cavity (the cavity of the stomach or intestine) opening by a mouth externally, at the upper pole of the longitudinal axis (opposite the basal point of fixture). The entire body is here in a certain sense a mere stomach or intestinal canal, as in the case of the Gastrula. The wall of the sack, which is both body wall and intestinal wall, consists of two layers or coats of cells, a fringed entoderm, or gastral layer (corresponding with the inner or vegetative germ-layer of the higher animals), and an unfringed exoderm or dermal layer (corresponding with the external or animal germ-layer of the higher animals). The original Protascus, a true likeness of which is still furnished by the Ascula, probably formed egg-cells and sperm-cells out of its gastral layer.

The Protascads—as we will call the most ancient group of vegetable animals, represented by the Protascus-type—divided into two lines or branches, the Spongiæ and the Sea-nettles, or Acalephæ. I have shown in my Monograph of the Calcareous Sponges (vol. i. p. 485) how closely these two main classes of Zoophytes are related, and how they must both be derived, as two diverging forms, from the Protascus-form. The primary form of Spongiæ, which Ihave there called Archispongia, arose out of the Protascus by the formation of pores through its body-wall; the primary form of Sea-nettles, which I there called Archydra, developed out of the Protascus by the formation of nettle-organs, as also by the formation of feelers or tentacles.

The main-class or branch of the Sponges, Spongiæ, or Porifera, lives in the sea, with the single exception of the green fresh-water Sponge (Spongilla). These animals were long considered as plants, later as Protista; in most Manuals they are still classed among the primæval animals, or Protozoa. But since I have demonstrated their development out of the Gastrula, and the construction of their bodies of two cellular germ-layers (as in all higher animals), their close relationship to Sea-nettles, and especially to the Hydrapolyps, seems finally to be established. The Olynthus especially, which I consider as the common primary form of calcareous sponges, has thrown a complete and unmistakable light upon this point.

The numerous forms comprised in the class of Spongiæ have as yet been but little examined; they may be divided into three legions and eight orders. The first legion consists of the soft, gelatinous Mucous Sponges (Myxospongiæ), which are characterized by the absence of any hard skeleton. Among them are, on the one hand, the long-since-extinct primary forms of the whole class, the type of which I consider to be the Archispongia; on the other hand there are the still living, gelatinous sponges, of which the Halisarca is best known. We can obtain a notion of the Archispongia, the most ancient primæval sponge, if we imagine the Olynthus (see Frontispiece), to be deprived of its radiating calcareous spiculæ.

The second legion of Spongiæ contains the Fibrous Sponges (Fibrospongiæ), the soft body of which is supported by a firm, fibrous skeleton. This fibrous skeleton often consists merely of so-called "horny fibres," formed of a very elastic, not readily destructible, organic substance. This is the case for instance in our common bathing Sponge (Euspongia officinalis), the purified skeleton of which we use every morning when washing. Blended with the horny, fibrous skeleton of many of these Sponges, there are numerous flinty spicula; this is the case for example with the fresh-water Sponge (Spongilla). In others the whole skeleton consists of only calcareous or silicious spicula which are frequently interwoven into an extremely beautiful lattice-work, as in the celebrated Venus' Flower Basket (Euplectella). Three orders of fibrous sponges may be distinguished according to the different formation of the spicula, namely, Chalynthina, Geodina, and Hexactinella. The natural history of the fibrous sponges is of especial interest to the Theory of Descent, as was first shown by Oscar Schmidt, the greatest authority on this group of animals. In no other group, perhaps, can the unlimited pliability of the specific form, and its relation to Adaptation and Inheritance, be so clearly followed step by step; perhaps in no other group is the species so difficult to limit and define.

This proposition, which applies to the great legion of the Fibrous Sponges, applies in a still higher degree to the smaller but exceedingly interesting legion of the calcareous sponges (Calcispongiæ), on which in 1872, after five years' careful examination, I published a comprehensive Monograph.

The sixty plates of figures accompanying this Monograph explain the extreme pliability of these small sponges "good species" of which, in fact, cannot be spoken of in the usual systematic sense. We find among them only varying series of forms, which do not even completely transmit their specific form to their nearest descendants, but by adaptation to subordinate, external conditions of existence, perpetually change. It frequently occurs here, that there arise out of one and the same stock different form-species, which according to the usual system would belong to several quite distinct genera; this is the case, for instance, with the remarkable Ascometra (Frontispiece, Fig. 10.) The entire external bodily form is much more pliable and protean in Calcareous Sponges than in the silicious sponges, which are characterized by possessing silicious spicula, forming a beautiful skeleton. Through the study of the comparative anatomy and ontogeny of calcareous sponges, we can recognise, with the greatest certainty, the common primary form of the whole group, namely, the sack-shaped Olynthus, whose development is represented in the Frontispiece (compare its explanation in the Appendix). Out of the Olynthus (Fig. 9 on the Frontispiece), the order of the Ascones was the first to develop, out of which, at a later period, the two other orders of Calcareous Sponges, the Leucones and Sycones, arose as diverging branches. Within these orders, the descent of the individual forms can again be followed step by step. Thus the Calcareous Sponges in every respect confirm the proposition which I have elsewhere maintained: that "the natural history of sponges forms a connected and striking argument in favour of Darwin."

The second main class or branch in the tribe of Zoophytes is formed by the Sea-nettles (Acalephæ, or Cnidæ). This interesting group of animals, so rich in forms, is composed of three different classes, namely, the Hood-jellies (Hydromedusæ), the Comb-jellies (Ctenophora), and the Corals (Coralla). The hypothetical, extinct Archydra must be looked upon as the common primary form of the whole group; it has left two near relations in the still living fresh-water polyps (Hydra and Cordylophora). The Archydra was very closely related to the simplest forms of Spongiæ (Archispongia and Olynthus), and probably differed from them only by possessing nettle organs, and by the absence of cutaneous pores. Out of the Archydra there first developed the different Hydroid polyps, some of which became the primary forms of Corals, others the primary forms of Hydromedusæ. The Ctenophora developed later out of a branch of the latter.

The Sea-nettles differ from the Spongiæ (with which they agree in the characteristic formation of the system of the alimentary canal) principally by the constant possession of nettle organs. These are small bladders filled with poison, large numbers—generally millions—of which are dispersed over the skin of the sea nettles, and which burst and empty their contents when touched. Small animals are killed by this; in larger animals this nettle poison causes a slight inflammation of the skin, just as does the poison of our common nettles. Any one who has often bathed in the sea, will probably have at times come in contact with large Hood-jellies (Jelly-fish), and become acquainted with the unpleasant burning feeling which their nettle organs can produce. The poison in the splendid blue Jelly-fish, Physalia, or Portuguese Man-of-war, acts so powerfully that it may lead to the death of a human being.

The class of Corals (Coralla) lives exclusively in the sea, and is more especially represented in the warm seas by an abundance of beautiful and highly-coloured forms like flowers. Hence they are also called Flower-animals (Anthozoa). Most of them are attached to the bottom of the sea, and contain an internal calcareous skeleton. Many of them by continued growth produce such immense stocks that their calcareous skeletons have formed the foundation of whole islands, as is the case with the celebrated coral reefs and atolls of the South Seas, the remarkable forms of which were first explained by Darwin.(13) In corals the counterparts, or antimera—that is, the corresponding divisions of the body which radiate from and surround the central main axis of the body—exist sometimes to the number of four, sometimes to the number of six or eight. According to this we distinguish three legions, the Fourfold (Tetracoralla), Sixfold (Hexacoralla), and Eightfold corals (Octocoralla). The fourfold corals form the common primary group of the class, out of which the sixfold and eightfold have developed as two diverging branches.

The second class of Sea-nettles is formed by the Hood-jellies (Medusæ) or Polyp-jellies (Hydromedusæ). While most corals form stocks like plants, and are attached to the bottom of the sea, the Hood-jellies generally swim about freely in the form of gelatinous bells. There are, however, numbers of them, especially the lower forms, which adhere to the bottom of the sea, and resemble pretty little trees. The lowest and simplest members of this class are the little fresh-water polyps (Hydra and Cordylophora). We may look upon them as but little changed descendants of those Primæval polyps (Archydræ), from which, during the primordial period, the whole division of the Sea-nettles originated. Scarcely distinguishable from the Hydra are the adherent Hydroid polyps (Campanularia, Tubularia), which produce freely swimming medusæ by budding, and out of the eggs of these there again arise adherent polyps. These freely swimming Hood-jellies are mostly of the form of a mushroom, or of an umbrella, from the rim of which many long and delicate tentacles hang. They are among the most beautiful and most interesting inhabitants of the sea. The remarkable history of their lives, and especially the complicated alternation of generation of polyps and medusæ, are among the strongest proofs of the truth of the theory of descent. For just as Medusæ still daily arise out of the Hydroids, did the freely swimming medusa-form originally proceed, phylogenetically, out of the adherent polyp-form. Equally important for the theory of descent is the remarkable division of labour of the individuals, which among some of them is developed to an astonishingly high degree, more especially in the splendid Siphonophora.(37) (Plate VII. Fig. 13).

The third class of Sea-nettles—the peculiar division of Comb-jellies (Ctenophora), probably developed out of a branch of the Hood-jellies. The Ctenophora, which are also called Ribbed-jellies, possess a body of the form of a cucumber, which, like the body of most Hood-jellies, is as clear and transparent as crystal or cut glass. Comb or Ribbed-jellies are characterized by their peculiar organs of motion, namely, by eight rows of paddling, ciliated leaflets, which run in the form of eight ribs from one end of the longitudinal axis (from the mouth) to the opposite end. Those with narrow mouths (Stenostoma) probably developed later out of those with wide mouths (Eurystoma). (Compare Plate VII. Fig. 16.)

The third tribe of the animal kingdom, the phylum of Worms or worm-like animals (Vermes, or Helminthes), contains a number of diverging branches. Some of these numerous branches have developed into well-marked and perfectly independent classes of Worms, but others changed long since into the original, radical forms of the four higher tribes of animals. Each of these four higher tribes (and likewise the tribe of Zoophytes) we may picture to ourselves in the form of a lofty tree, whose branches represent the different classes, orders, families, etc. The phylum of Worms, on the other hand, we have to conceive as a low bush or shrub, out of whose root a mass of independent branches shoot up in different directions. From this densely branched shrub, most of the branches of which are dead, there rise four high stems with many branches. These are the four lofty trees just mentioned as representing the higher phyla—the Echinoderma, Articulata, Mollusca, and Vertebrata. These four stems are directly connected with one another at the root only, to wit, by the common primary group of the Worm tribe.

The extraordinary difficulties which the systematic arrangement of Worms presents, for this reason merely, are still more increased by the fact that we do not possess any fossil remains of them. Most of the Worms had and still have such soft bodies that they could not leave any characteristic traces in the neptunic strata of the earth. Hence in this case again we are entirely confined to the records of creation furnished by ontogeny and comparative anatomy. In making then the exceedingly difficult attempt to throw a few hypothetical rays of light upon the obscurity of the pedigree of Worms, I must therefore expressly remark that this sketch, like all similar attempts possesses only a provisional value.

The numerous classes distinguished in the tribe of Worms, and which almost every zoologist groups and defines according to his own personal views, are, in the first place, divided into two essentially different groups or branches, which in my Monograph of the Calcareous Sponges I have termed Acœlomi and Cœlomati. For all the lower Worms which are comprised in the class of Flat-worms (Platyhelminthes), (the Gliding-worms, Sucker-worms, Tape-worms), differ very strikingly from other Worms, in the fact that they possess neither blood nor body-cavity (no cœlome); they are, therefore, called Acœlomi. The true cavity, or cœlome, is completely absent in them as in all the Zoophytes; in this important respect the two groups are directly allied. But all other Worms (like the four higher tribes of animals) possess a genuine body-cavity and a vascular system connected with it, which is filled with blood; hence we class them together as Cœlomati.

The main division of Bloodless Worms (Acœlomi) contains, according to our phylogenetic views, besides the still living Flat-worms, the unknown and extinct primary forms of the whole tribe of Worms, which we shall call the Primæval Worms (Archelminthes). The type of these Primæval Worms, the ancient Prothelmis, may be directly derived from the Gastræa (p. 133). Even at present the Gastrula-form—the faithful historical portrait of the Gastræa—recurs in the ontogenesis of the most different kinds of worms as a transient larva-form. The ciliated Gliding-worms (Turbellaria), the primary group of the present Planary or Flat-worms (Platyhelminthes), are the nearest akin to the Primæval Worms. The parasitical Sucker-worms (Trematoda) arose out of the Gliding-worms, which live freely in water, by adaptation to a parasitical mode of life; and out of them later on—by an increasing parasitism—arose the Tape-worms (Cestoda).

Out of a branch of the Acœlomi arose the second main division of the Worm tribe, the Worms with blood and body-cavity (Cœlomati): of these there are seven different classes.

The Pedigree on p. 151 shows how the obscure phylogeny of the seven classes of Cœlomati may be supposed to stand. We shall, however, mention these classes here quite briefly, as their relationships and derivation are, at present, still very complicated and obscure. More numerous and more accurate investigations of the ontogeny of the different Cœlomati will at some future time throw light upon their phylogenesis.

The Round Worms (Nemathelminthes) which we mention as the first class of the Cœlomati, and which are characterized by their cylindrical form, consist principally of parasitical Worms which live in the interior of other animals. Of human parasites, the celebrated Trichinæ, the Maw-worms, Whip-worms, etc., for example, belong to them. The Star-worms (Gephyrea) which live exclusively in the sea are allied to round worms, and the comprehensive class of Ring-worms (Annelida) are allied to the former. To the Ring-worms, whose long body is composed of a number of segments, all alike in structure, belong the Leeches (Hirudinea), Earth-worms (Lumbricina), and all the marine bristle-footed Worms (Chætopoda).

Nearly akin to them are the Snout-worms (Rhynchocœla), and the small microscopic Wheel-worms (Rotifera). The unknown, extinct, primary forms of the tribe of Sea-stars (Echinoderma), and of the tribe of the articulated animals (Arthropoda), were nearest akin to the Ring-worms. On the other hand, we must probably look for the primary forms of the great tribe of Molluscs in extinct Worms, which were very closely related to the Moss-polyps (Bryozoa) of the present day; and for the primary forms of the Vertebrata in the unknown Cœlomati, whose nearest kin of the present day are the Sea-sacs, especially the Ascidia.

The class of Sea-sacs (Tunicata) is one of the most remarkable among Worms. They all live in the ocean, where some of the Ascidiæ adhere to the bottom, while others (the sea-barrels, or Thaliacea) swim about freely. In all of them the non-jointed body has the form of a simple barrel-shaped sack, which is surrounded by a thick cartilaginous mantle. This mantle consists of the same non-nitrogenous combination of carbon, which, under the name of cellulose, plays an important part in the Vegetable Kingdom, and forms the largest portion of vegetable cellular membranes, and consequently also the greater part of wood. The barrel-shaped body generally possesses no external appendages. No one would recognise in them a trace of relationship to the highly differentiated vertebrate animals. And yet this can no longer be doubted, since Kowalewsky's investigations, which in the year 1867 suddenly threw an exceedingly surprising and unmistakable light upon them. From these investigations it has become clear that the individual development of the adherent simple Ascidian Phallusia agrees in most points with that of the lowest vertebrate animal, namely, the Lancelet (Amphioxus lanceolatus). The early stages of the Ascidia possess the beginnings of the spinal marrow and the spinal column (chorda dorsalis) lying beneath it, which are the two most essential and most characteristic organs of the vertebrate animal. Accordingly, of all invertebrate animals known to us, the Tunicates are without doubt the nearest blood relations of the Vertebrates, and must be considered as the nearest relations of those Worms out of which the vertebrate tribe has developed. (Compare Plates XII. and XIII.)

While thus different branches of the Cœlomatous group of the Worms furnish us with several genealogical links leading to the four higher tribes of animals, and give us important phylogenetic indications of their origin, the lower group of Acœlomi, on the other hand, show close relationships to the Zoophytes, and to the Primæval animals. The great phylogenetic interest of the Worm tribe rests upon this peculiar intermediate position.

PEDIGREE AND HISTORY OF THE ANIMAL KINGDOM

II. Mollusca, Star-fishes, and Articulated Animals.

The great natural main groups of the animal kingdom, which we have distinguished as TRIBES, or PHYLA ("types" according to Bär and Cuvier), are not all of equal systematic importance for our phylogeny or history of the pedigree of the living world. They can neither be classed in a single series of stages, one above another, nor be considered as entirely independent stems, nor as equal branches of a single family-tree. It seems rather (as we saw in the last chapter) that the tribe of Protozoa, the so-called primæval animals, is the common radical group of the whole animal kingdom. Out of the Gastræada—which we class among the Protozoa— the Zoophytes and the Worms have developed, as two diverging branches. We must now in turn look upon the varied and much-branching tribe of Worms as the common primary group, out of which (from perfectly distinct branches) arose the remaining tribes, the four higher phyla of the animal kingdom. (Compare the Pedigree, p. 133.)

Let us now take a genealogical look at these four higher tribes of animals, and try whether we cannot make out the most important outlines of their pedigree. Even should this attempt prove defective and imperfect, we shall at all events have made a beginning, and paved the road for subsequent and more satisfactory attempts.

It does not matter in what succession we take up the examination of the four higher tribes. For these four phyla have no close relationship whatever among one another, but have grown out from entirely distinct branches of the group of Worms (p. 133). We may consider the tribe of Molluscs as the most imperfect and the lowest in point of morphological development. We nowhere meet among them with the characteristic articulation or segmented formation of the body, which distinguishes even the Ring-worms, and which in the other three higher tribes—the Echinoderma, Articulata, and Vertebrata—is most essentially connected with the high development of their forms, their differentiation, and perfection. The body in all Molluscs—in mussels, snails, etc.—is a simple non-jointed sack, in the cavity of which lie the intestines. The nervous system consists not of a cord but of several distinct (generally three) pairs of knots loosely connected with one another. For these and many other anatomical reasons, I consider the tribe of Molluscs (in spite of the high physiological development of its most perfect forms) to be morphologically the lowest among the four higher tribes of animals.

Whilst, for reasons already given, we exclude the Moss-polyps, and Tunicates—which have hitherto been generally classed with the tribe of Molluscs—we retain as genuine Molluscs the following four classes: Lamp-shells, Mussels, Snails, and Cuttles. The two lower classes of Molluscs, the Lamp-shells and Mussels, possess neither head nor teeth, and they can therefore be comprised under one main class, or branch, as headless animals (Acephala), or toothless animals (Anodontoda). This branch is also frequently called that of the clam-shells (Conchifera, or Bivalvia), because all its members possess a two-valved calcareous shell. In contrast to these the two higher classes of Molluscs, the snails and cuttles, may be represented as a second branch with the name of Head-bearers (Cephalophora), or Tooth-bearers (Odontophora), because both head and teeth are developed in them.

The soft, sack-shaped body in most Molluscs is protected by a calcareous shell or house, which in the Acephala (lamp-shells and mussels) consists of two valves, but in the Cephalophora (snails and cuttles) is generally a spiral tube (the so-called snail's house). Although these hard skeletons are found in large quantities in a petrified state in all the neptunic strata, yet they tell us but little of the historical development of the tribe, which must have taken place for the most part in the primordial period. Even in the Silurian strata we find fossil remains of all the four classes of Molluscs, one beside the other, and this, conjointly with much other evidence, distinctly proves that the tribe of Molluscs had then obtained a strong development, when the higher tribes, especially the Articulates and Vertebrates, had scarcely got beyond the beginning of their historical development. In subsequent periods, especially in the primary and secondary periods, these higher tribes increased in importance more and more at the expense of Molluscs and Worms, which were no match for them in the struggle for life, and accordingly decreased in number. The still living Molluscs and Worms must be considered as only a proportionately small remnant of the vast molluscan fauna, which greatly predominated in the primordial and primary periods over the other tribes. (Compare Plate VI. and explanation in the Appendix.)

No tribe of animals shows more distinctly than do the Molluscs, how very different the value of fossils is in geology and in phylogeny. In geology the different species of the fossil shells of Molluscs are of the greatest importance because they serve as excellent marks whereby to characterize the different groups of strata, and to fix their relative ages. As far as relates to the genealogy of Molluscs, however, they are of very little value, because, on the one hand, the shells are parts of quite subordinate morphological importance, and because the actual development of the tribe belongs to the earlier primordial period, from which no distinct fossils have been preserved. If therefore we wish to construct the pedigree of Molluscs, we are mainly dependent upon the records of ontogeny and comparative anatomy from which we obtain something like the following result. (Gen. Morph. ii. Plate VI. pp. 102-116.)

The lowest stage of the four classes of genuine Molluscs known to us, is occupied by the Lamp-shells or Spiral-gills (Spirobranchia), frequently but inappropriately called Arm-footers (Brachiopoda), which have become attached to the bottom of the sea. There now exist but few forms of this class; for instance, some species of Lingula, Terebratula, and others akin to them, which are but feeble remnants of the great variety of forms which represented the Lamp-shells in earlier periods of the earth's history. In the Silurian period they constituted the principal portion of the whole Mollusc tribe. From the agreement which, in many respects, their early stage of development presents with the Moss animals, it has been concluded that they have developed out of Worms, which were nearly related to this class. Of the two sub-classes of Lamp-shells, the Hinge-less (Ecardines) must be looked upon as the lower and more imperfect, the Hinged (Testicardines) as the higher and more fully developed group.

The anatomical difference between the Lamp-shells and the three other classes of Molluscs is so considerable that the latter may be distinguished from the former by the name of Otocardia. All the Otocardia have a heart with chamber (ventricle) and ante-chamber (auricle), whereas Lamp-shells do not possess the ante-chamber. Moreover, the central nervous system is developed only in the former (and not in the latter) in the shape of a complete pharyngeal ring. Hence the four classes of Molluscs may be grouped in the following manner:—

I. Molluscs

without head.

Acephala.

1. Lamp-shells

(Spirobranchia)

I. Haplocardia

(with simple heart)
2. Mussels
(Lamellibranchia)
II. Otocardia
(with chamber and ante-chamber to the heart)
I. Molluscs with head.
Cephalophora.
3. Snails
(Cochlides).
4. cuttles
cephalopoda.

The result of these structural dispositions for the history of the pedigree of Molluscs, which is confirmed by palæontology, is that Lamp-shells stand much nearer to the primæval root of the whole tribe of Molluscs than do the Otocardia. Probably Mussels and Snails developed as two diverging branches out of Molluscs, which were nearly akin to the Lamp-shells.

Mussels, or Plate-gills (Lamellibranchia), possess a bivalved shell like the Lamp-shells. In the latter, one of the two valves covers the back, the other the belly of the animal; whereas in Mussels the two valves lie symmetrically on the right and left side of the body. Most Mussels live in the sea, only a few in fresh water. The class is divided into two sub-classes, Asiphonia and Siphonida, of which the latter were developed at a later period out of the former. Among the Asiphonia are Oysters, mother-of-pearl Shells, and fresh water Mussels; among the Siphonida, which are characterized by a respiratory tube, are the Venus-shells, Razor-shells, and Burrowing Clams. The higher Molluscs seem to have developed at a later period out of those without head and teeth; they are distinguished from the latter by the distinct formation of the head, and more especially by a peculiar kind of tooth apparatus. Their tongue presents a curious plate, armed with a great number of teeth. In our common Vineyard Snail (Helix pomatia) the number of teeth amount to 21,000, and in the large Garden Slug (Limax maximus) to 26,800.

We distinguish two sub-classes among the Snails (Cochlides, or Gasteropoda), namely, the Stump-headed and the Large-headed Snails. The Stump-headed Snails (Perocephala) are very closely allied to Mussels (through the Tooth-shells), and also to the Cuttle-fish (through the Butterfly-snails). The more highly developed Snails, with large heads (Delocephala), can be divided into Snails with gills (Branchiata) and Snails with lungs (Pulmonata). Among the latter are the Land-snails, the only Molluscs which have left the water and become habituated to a life on land. The great majority of Snails live in the sea, only a few live in fresh water. Some River-snails in the tropics (the Ampullaria) are amphibious, living sometimes on land, sometimes in water, and at one time they breathe through gills, at another through lungs. They have both kinds of respiratory organs, like the Mud-fish and Gilled Newts among the Vertebrata.

The fourth and last class, and at the same time the most highly developed class of Molluscs, is that of the Cuttles, or Poulps, also called Cephalopoda (foot attached to the head). They all live in the sea, and are distinguished from Snails by eight, ten, or more long arms, which surround the mouth in a circle. The Cuttles existing in our recent oceans—the Sepia, Calamary, Argonaut, and Pearly Nautilus—are, like the few Spiral-gill Lamp-shells of the present time, but a poor remnant of the host which represents this class in the oceans of the primordial, primary, and secondary periods. The numerous fossil "Ammon's horns" (Ammonites), "pearl boats" (Nautilus), and "thunderbolts" (Belemnites) are evidences of the long since extinct splendour of the tribe. The Poulps, or Cuttles, have probably developed out of a low branch of the snail class, out of the Butterfly-snails (Pteropoda) or kindred forms.

The different sub-classes and orders, distinguished in the four classes of Molluscs, whose systematic succession is given on the Table (p. 160), furnish various proofs of the validity of the law of progress by their historical development and by the systematic development corresponding to it. As however these subordinate groups of Molluscs are in themselves of no further special interest, I must refer to the sketch of their pedigree on p. 161, and to the detailed pedigree of Molluscs which I have given in my General Morphology, and I shall now at once turn to the consideration of the tribe of Star-fishes.

The Star-fishes (Echinoderma, or Estrellæ) among which are the four classes of Sea-stars, Sea-lilies, Sea-urchins, and Sea-cucumbers are one of the most interesting divisions of the animal kingdom, and yet we know less about them than about any. They all live in the sea. Every one who has been at the sea shore must have seen at least two of their forms, the Sea-stars and the Sea-urchins. The tribe of Star-fishes must be considered as a completely independent tribe of the animal kingdom on account of its very peculiar organization, and must be carefully distinguished from the Animal-plants—Zoophytes, or Cœlenterata, with which it is still frequently but erroneously classed under the name Radiata (as for example, by Agassiz, who even to this day defends this error of Cuvier's, together with many others).

All Echinoderma are characterized, and at the same time distinguished from all other animals, by a very remarkable apparatus for locomotion, which consists of a complicated system of canals or tubes, filled with sea water from without. The sea water in these aqueducts is moved partly by the strokes of the cilia, or vibratile hairs lining their walls, and partly by the contractions of the muscular walls of the tubes themselves, which resemble india-rubber bags. The water is pressed from the tubes into a number of little hollow feet, which thereby become widely distended, and are then employed for walking and suction. The Sea-stars are moreover characterized by a peculiar calcareous formation in the skin, which in most cases forms a firm, well-closed coat of mail, composed of a number of plates. In almost all Echinoderma the body consists of five radii (counterparts, or antimera) standing round the main axis of the body, where they meet. It is only in some species of Sea-stars that the number of these radii amount to more than five—to 6-9, 10-12, or even to 20-40; and in this case the number of radii is generally not constant, but varies in different individuals of one species.

The historical development and the pedigree of the Echinoderma are completely revealed to us by their numerous and, in most cases, excellently preserved fossil remains, by their very remarkable individual developmental history, and by their interesting comparative anatomy; this is the case with no other tribe of animals, even the Vertebrata themselves are not to be excepted. By a critical use of those three archives, and by a careful comparison of the results derived from their study, we obtain the following genealogy of the Star-fishes, which I have already published in my General Morphology (vol. ii. Plate IV. pp. 62-77.)

The most ancient and original group of the Star-fishes, the primary form of the whole phylum, consists of the class of the true Sea-stars (Asterida). This is established by numerous and important arguments in anatomy and the history of development, but above all by the irregular and varying number of the radii, or antimera, which in all other Echinoderma is limited, without exception, to five. Every Star-fish consists of a central, small, body-disc, all round the circumference of which are attached five or several long articulated arms. Each arm of the Star-fish essentially corresponds in its organisation with an articulated worm of the class of Ring-worms, or Annelida (p. 149). I therefore consider the Star-fish as a genuine stock or cormus of five or more articulated worms, which have arisen by the star-wise growth of a number of buds out of a central mother-worm. The connected members, thus grouped like the rays of a star, have inherited from the mother-worm the common opening of the mouth, and the common digestive cavity (stomach) lying in the central body-disc. The end by which they have grown together, and which fuses in the common central disc, probably corresponds to the posterior end of the original independent worms.

In exactly the same way several individuals of certain kinds of worms are united so as to form a star-like cormus. This is the case in the Botryllidæ, compound Ascidians, belonging to the class of the Tunicata. Here also the posterior ends of the individual worms have grown together, and have formed a common outlet for discharges, a central cloaca; whereas at the anterior end each worm still possesses its own mouth. In Star-fishes the original mouths have probably become closed in the course of the historical development of the cormus, or colony, whereas the cloaca has developed into a common mouth for the whole cormus.

Hence the Star-fishes would be compound stocks of worms which, by the radial formation of buds, have developed out of true articulated worms, or Annelids. This hypothesis is most strongly supported by the comparative anatomy, and by the ontogeny of some Star-fishes (Colastra), and of segmented worms. The many-jointed Ring-worms (Annelida) in their inner structure are closely allied to the individual arms or radii of the Star-fishes, that is to the original single worms, which each arm represents. Each of the five worms of the Star-fish is a chain composed of a great number of equi-formal members, or metamera, lying one behind the other, like every segmented Worm, and every Arthropod. As in the latter a central nervous cord, the ventral nerve cord runs along the central line of the ventral wall of each segment. On each metameron there is a pair of non-jointed feet, and besides these, in most cases, one or more hard thorns or bristles similar to those of many Ring-worms. A detached arm of a Star-fish can lead an independent life, and can then, by the radially-directed growth of buds at one end, again become a complete star.

The most important proofs, however, of the truth of my hypothesis are furnished by the ontogeny or the individual development of the Echinoderma. The most remarkable facts of this ontogeny were first discovered in the year 1848 by the great zoologist, Johannes Müller of Berlin. Some of its most important stages are represented on Plates VIII. and IX. (Compare their explanation in the Appendix.) Fig. A on Plate IX. shows us a common Sea-star (Uraster), Fig. B, a Sea-lily (Comatula), Fig. C, a Sea-urchin (Echinus), and Fig. D, a Sea-cucumber (Synapta). In spite of the extraordinary difference of form manifested by these four representatives of the different classes of Star-fishes, yet the beginning of their development is identical in all cases. Out of the egg an animal-form develops which is utterly different from the fully developed Star-fish, but very like the ciliated larvæ of certain segmented Worms (Star-worms and Ring-worms). This peculiar animal-form is generally called the "larva," but more correctly the "nurse" of these Star-fish. It is very small and transparent, swims about by means of a fringe of cilia, and is always composed of two equal symmetrical halves or sides. The fully grown Echinoderm, however—which is frequently more than a hundred times larger, and quite opaque—creeps at the bottom of the sea, and is always composed of at least five co-ordinate pieces, or antimera, in the form of radii. Plate VIII. shows the development of the "nurses" of the four Echinoderms represented on Plate IX.

The fully developed Echinoderm arises by a very remarkable process of budding in the interior of the "nurse," of which it retains little more than the stomach. The nurse, erroneously called the "larva," of the Echinoderm, must accordingly be regarded as a solitary worm, which by internal budding produces a second generation, in the form of a stock of star-shaped and connected worms. The whole of this process is a genuine alternation of generations, or metagenesis, not a "metamorphosis," as is generally though erroneously stated. A similar alternation of generations also occurs in many other worms, especially in some star worms (Sipunculidæ), and cord worms (Nemertinæ). Now if, bearing in mind the fundamental law of biogeny, we refer the ontogeny of Echinoderma to their phylogeny, then the whole historical development of the Star-fishes suddenly becomes clear and intelligible to us, whereas without this hypothesis it remains an insoluble mystery. (Compare Gen. Morph. ii pp. 95-99.)

Besides the reasons mentioned, there are many other facts (principally from the comparative anatomy of Echinoderma) which most distinctly prove the correctness of my hypothesis. I established this hypothesis in 1866, without having any idea that fossil articulated worms still existed, apparently answering to the hypothetical primary forms. Such have in the mean time, however, really been discovered. In a treatise "On the Equivalent of the North American Taconic Schist in Germany,"3 Geinitz and Liebe, in 1867, have described a number of articulated Silurian worms, which completely confirm my suppositions. Numbers of these very remarkable worms are found in an excellent state of preservation in the slates of Würzbach, in the upper districts of Reusz. They are of the same structure as the articulated arm of a Star-fish, and evidently possessed a hard coat of mail, a much denser, more solid cutaneous skeleton than other worms in general. The number of body-segments, or metamera, is very considerable, so that the worms, although no more than a quarter or half an inch in breadth, attained a length of from two to three feet. The excellently preserved impressions, especially those of the Phyllodocites thuringiacus and Crossopodia Henrici, are so like the arms of many Star-fish (Colastra) that their true blood relationship seems very probable. This primæval group of worms, which are most probably the ancestors of Star-fish, I call Mailed worms (Phracthelminthes, p. 150.)

The three other classes of Echinoderma evidently arose at a later period out of the class of Sea-stars which have most faithfully retained the original form of the stellate colony of worms. The Sea-lilies, or Crinoida, differ least from them, but having given up the free, slow motion possessed by other Sea-stars, they have become adherent to rocks, etc., and form for themselves a long stalk. Some Encrinites, however (for example, the Comatulæ, Fig. B, on Plates VIII. and IX.), afterwards detach themselves from their stalk. The original worm individuals in the Crinoida are indeed no longer preserved in the same independent condition as in the case of the common star-fish; but they nevertheless always possess articulated arms extending from a common central disc. Hence we may unite the Sea-lilies and Sea-stars into a main-class, or branch, characterized as possessing articulated arms (Colobrachia).

In the other two classes of Echinoderma, the Sea-urchins and Sea-cucumbers, the articulated arms are no longer present as independent parts, but, by the increased centralization of the stock, have completely fused so as to form a common, inflated, central disc, which now looks like a simple box or capsule without arms. The original stock of five individuals has apparently degenerated to the form-value of a simple individual, a single person. Hence we may represent these two classes as a branch characterized as being without arms (Lipobrachia), equivalent to those which possess articulated arms. The first of these two classes, that of Sea-urchins (Echinida) takes its name from the numerous and frequently very large thorns which cover the hard shell, which is itself artistically built up of calcareous plates. (Fig. C, Plates VIII. and IX.) The fundamental form of the shell itself is a pentagonal pyramid. The Sea-urchins probably developed directly out of the group of Sea-stars. The different classes and orders of marine lilies and stars which are given in the following table, illustrate the laws of progress and differentiation in a striking manner. In each succeeding period of the earth's history we see the individual classes continually increasing in variety and perfection. (Gen. Morph. ii. Plate IV.)

The history of three of these classes of Star-fish is very minutely recorded by numerous and excellently preserved fossils, but on the other hand, we know almost nothing of the historical development of the fourth class, that of the Sea-cucumbers (Holothuriæ). These curious sausage-shaped Star-fish manifest externally a deceptive similarity to worms. (Fig. D, Plates VIII. and IX.) The skeletal structures in their skin are very imperfect, and hence no distinct remains of their elongated, cylindrical, worm-like body could be preserved in a fossil state. However, from the comparative anatomy of the Holothuriæ, we can infer that they have arisen, by the softening of the cutaneous skeleton, from members of the class of Sea-urchins.

From the Star-fish we turn to the fifth and most highly developed tribe of the invertebrate animals, namely, the phylum of Articulata, or those with jointed feet (Arthropoda). As has already been remarked, this tribe corresponds to Linnæus' class of Insects. It contains four classes: (1) the genuine six-legged Insects, or Flies; (2) the eight-legged Spiders; (3) the Centipedes, with numerous pairs of legs; and (4) the Crabs, or Crustacea, whose legs vary in number. The last class breathe water through gills, and may therefore be contrasted as the main-class of gill-breathing Arthropoda, or Gilled Insects (Carides), with the three first classes. The latter breathe air by means of peculiar wind-pipes, or tracheæ, and may therefore appropriately be united to form the main-class of the trachea-breathing Arthropoda, or Tracheate Insects (Tracheata).

In all animals with articulated feet, as the name indicates, the legs are distinctly articulated, and by this, as well as by the strong differentiation of the separate parts of the body, or metamera, they are sharply distinguished from Ringed worms, with which Bär and Cuvier classed them. They are, however, in every respect so like the Ringed worms that they can scarcely be considered altogether distinct from them. They, like the Ringed worms, possess a very characteristic form of the central nervous system, the so-called ventral marrow, which commences in a gullet-ring encircling the mouth. From other facts also, it is evident that the Arthropoda developed at a late period out of articulated worms. Probably either the Wheel Animalcules or the Ringed worms are their nearest blood relations in the Worm tribe. (Gen. Morph. ii. Plate V. pp. 85-102.)

Now, although the derivation of the Arthropoda from ringed Worms may be considered as certain, still it cannot with equal assurance be maintained that the whole tribe of the former has arisen out of one branch of the latter. For several reasons seem to support the supposition that the Gilled Arthropods have developed out of a branch of articulated worms, different from that which gave rise to the Tracheate Arthropods. But on the whole it remains more probable that both main-classes have arisen out of one and the same group of Worms. In this case the Tracheate Insects—Spiders, Flies, and Centipedes—must have branched off at a later period from the gill-breathing Insects, or Crustacea.

The pedigree of the Arthropoda can on the whole be clearly made out from the palæontology, comparative anatomy, and ontogeny of its four classes, although here, as everywhere else, many details remain very obscure. Not until the history of the individual development of all the different groups has become more accurately known than it is at present, can this obscurity be removed. The history of the class of Gilled Insects, or Crabs (Carides), is at present that best known to us; they are also called encrusted animals (Crustacea), on account of the hard crust or covering of their body. The ontogeny of these animals is extremely interesting and, like that of Vertebrate animals, distinctly reveals the essential outlines of the history of their tribe, that is, their phylogeny. Fritz Müller, in his work, "Für Darwin,"(16) which has already been referred to, has explained this remarkable series of facts in a very able manner.

The common primary form of all Crabs, which in most cases is even now the first to develop out of the egg, is originally one and the same, the so-called Nauplius. This remarkable primæval crab represents a very simple form of articulated animal, the body of which in general has the form of a roundish, oval, or pear-shaped disc, and has on its ventral side only three pairs of legs. The first of these is uncloven, the two subsequent pairs are forked. In front, above the mouth, lies a simple, single eye. Although the different orders of the Crustacean class differ very widely from one another in the structure of their body and its appendages, yet the early Nauplius form always remains essentially the same. In order to be convinced of this, let the reader look attentively at Plates X. and XI., a more detailed explanation of which is given in the Appendix. On Plate XI. we see the fully developed representatives of six different orders of Crabs, a Leaf-footed Crab (Limnetis, Fig. A c); a Stalked Crab (Lepas, Fig. D c); a Root Crab, (Sacculina, Fig. E c); a Boatman Crab (Cyclops, Fig. B c); a Fish Louse (Lernæocera, Fig. C c); and, lastly, a highly developed Shrimp (Peneus, Fig. F c). These six crabs vary very much, as we see, in the entire form of body, in the number and formation of the legs, etc. When, however, we look at the earliest stages, or "nauplius," of these six different classes, after they have crept out of the egg—those marked with corresponding letters on Plate X. (Fig. A n-F n)—we shall be surprised to find how much they agree. The different forms of Nauplius of these six orders differ no more from one another than would six different "good species" of one genus. Consequently, we may with assurance infer a common derivation of all those orders from a common Primæval Crab, which was essentially like the Nauplius of the present day.

The pedigree will show how we may at present approximately conceive the derivation of the twenty orders of Crustacea enumerated on p. 176, from the common primary form of the Nauplius. Out of the Nauplius form—which originally existed as an independent genus—the five legions of lower Crabs developed as diverging branches in different directions, which in the systematic survey of the class are united as Segmented Crabs (Entomostraca). The higher division of Mailed Crabs (Malacostraca) have likewise originated out of the common Nauplius form. The Nebalia is still a direct form of transition from the Phyllopods to the Schizopods, that is, to the primary form of the stalk-eyed and sessile-eyed Mailed Crabs. The Nauplius at this stage gives rise to another larva form, the so-called Zoëa, which is of great importance. The order of Schizopoda, those with cloven feet (Mysis, etc.), probably originated from this curious Zoëa; they are at present still directly allied, through the Nebalia to the Phyllopoda, those with foliaceous feet. But of all living crabs the Phyllopods are the most closely allied to the original primary form of the Nauplius. Out of the Schizopoda the stalk-eyed and sessile-eyed Mailed Crabs, or Malacostraca, developed as two diverging branches in different directions: the former through shrimps (Peneus, etc.), the latter through the Cumacea (Cuma, etc.), which are still living and closely allied to the Schizopoda. Among those with stalked eyes is the river crab (cray-fish), the lobster, and the others with long tails, or the Macrura, out of which, in the chalk period, the short-tailed crabs, or Brachyura, developed by the degeneration of the tail. Those with sessile eyes divide into the two branches of Flea-crabs (Amphipoda) and Louse-crabs (Isopoda); among the latter are our common

Rock-slaters and Wood-lice.

The second main-class of Articulated animals, that of the Tracheata, or air-breathing Tracheate Insects4 (Spiders, Centipedes, and Flies) did not develop until the beginning of the palæolithic era, after the close of the archilithic period, because all these animals (in contrast with the aquatic crabs) are originally inhabitants of land. It is evident that the Tracheata can have developed only after the lapse of the Silurian period when terrestrial life first began. But as fossil remains of spiders and insects have been found, even in the carboniferous beds, we can pretty accurately determine the time of their origin. The development of the first Tracheate Insects out of gill-bearing Zoëa-crabs, must have taken place between the end of the Silurian and the beginning of the coal period, that is, in the Devonian period.

Gegenbaur, in his excellent "Outlines of Comparative Anatomy,"(21) has lately endeavoured to explain the origin of the Tracheata by an ingenious hypothesis. The system of tracheæ, or air pipes, and the modifications of organization dependent upon it, distinguish Flies, Centipedes, and Spiders so much from other animals, that the conception of its first origin presents no inconsiderable difficulties to phylogeny. According to Gegenbaur, of all living Tracheate Insects, the Primæval Flies, or Archiptera, are most closely allied to the common primary form of the Tracheata. These insects—among which we may especially mention the delicate Day flies (Ephemera), and the agile dragon-flies (Libellula)—in their earliest youth, as larvæ, frequently possess external tracheate gills which lie in two rows on the back of the body, and are shaped like a leaf or paint-brush. Similar leaf or paint-brush shaped organs are met with as real water-breathing organs or gills, in many crabs and ringed worms, and, moreover, in the latter as real dorsal appendages or limbs. The "tracheate gills," found in the larvæ of many primæval winged insects, must in all probability be explained as "dorsal limbs," and as having developed out of the corresponding appendages of the Annelida, or possibly as having really arisen out of similar parts in Crustacea long since extinct. The present tracheal respiration of the Tracheata developed at a later period out of respiration through the "tracheate gills." The tracheate gills themselves, however, have in some cases disappeared, and in others become transformed into the wings of the Flies. They have disappeared entirely in the classes of Spiders and Centipedes, and these groups must accordingly be conceived of as degenerated or peculiarly developed lateral branches of the Fly class, which at an early period branched off from the common primary form of Flies; Spiders probably did so at an earlier period than Centipedes. Whether that common primary form of all Tracheata, which in my General Morphology I have named Protracheata, did develop directly out of genuine Ringed worms, or at first out of Crustacea of the Zoëa form (Zoëpoda, p. 212) will probably be settled at some future time by a more accurate knowledge and comparison of the ontogeny of the Tracheata, Crustacea, and Annelida. However, the root of the Tracheata, as well as that of the Crustacea, must in any case be looked for in the group of Ringed worms.

The genuine Spiders (Arachnida) are distinguished from Flies by the absence of wings, and by four pairs of legs; but, as is distinctly seen in the Scorpion-spiders and Tarantulæ, they, like Flies, possess in reality only three pairs of genuine legs. The apparent "fourth pair of legs" in spiders (the foremost) are in reality a pair of feelers. Among the still existing Spiders, there is a small group which is probably very closely allied to the common primary form of the whole class; this is the order of Scorpion-spiders, or Solifugæ, (Solpuga, Galeodes), of which several large species live in Africa and Asia, and are dreaded on account of their poisonous bite. Their body consists—as we suppose to have been the case in the common ancestor of the Tracheata—of a head possessing several pairs of feelers like legs, of a thorax, to the three rings of which are attached three pairs of legs, and of a hinder, body, or abdomen, consisting of many distinct rings. In the articulation of their body, the Solifugæ are therefore in reality more closely related to flies than to other spiders. Out of the Devonian Primæval Spiders, which were nearly related to the Solifugæ of the present day, the Long Spiders, the Tailor Spiders, and the Round Spiders probably developed as three diverging branches.

The Long Spiders (Arthrogastres), in which the earlier articulation of body has been better preserved than in Round Spiders, appear to be the older and more original forms. The most important members of this sub-class are the scorpions, which are connected with the Solifugæ through the Tarantella (or Phrynidæ). The small book scorpions, which inhabit our libraries and herbariums, appear as a degenerate lateral branch from the true scorpions. Mid-way between the Scorpions and Round Spiders are the long-legged Tailor-spiders (Opiliones) which have possibly arisen out of a special branch of the Solifugæ. The Pycnogonida, or No-body Crabs, and the Arctisca, or Bear Worms—still generally included among Long Spiders—must be completely excluded from the class of Spiders; the former belong to the Crustacea, the latter to Ringed worms.

Fossil remains of Long Spiders are found in the Coal. The second sub-class of the Arachnida, the Round Spiders (Sphærogastres), first appear in the fossil state in the Jura, that is, at a very much later period. They have developed out of a branch of the Solifuga, by the rings of the body becoming more and more united with one another. In the true Spinning Spiders (Araneæ), which we admire on account of their delicate skill in weaving, the union of the joints of the trunk, or metamera, goes so far, that the trunk now consists of only two pieces, of a head-breast (cephalo-thorax) with jaws, feelers, and four pairs of legs, and of a hinder body without appendages, where the spinning warts are placed. In Mites (Acarida), which have probably arisen by degeneration (especially by parasitism) out of a lateral branch of Spinning Spiders, even these two trunk pieces have become united and now form an unsegmented mass.

The class of Scolopendria, Myriapoda, or Centipedes, the smallest and poorest in forms of the four classes of Arthropoda, is characterized by a very elongated body, like that of a segmented Ringed worm, and often possesses more than a hundred pairs of legs. But these animals also originally developed out of a six-legged form of Tracheata, as is distinctly proved by the individual development of the millipede in the egg. Their embryos have at first only three pairs of legs, like genuine insects, and only at a later period do the posterior pairs of legs bud, one by one, from the growing rings of the hinder body. Of the two orders of Centipedes (which in our country live under barks of trees, in moss, etc.) the round, double-footed ones (Diplopoda) probably did not develop until a later period out of the older flat, single-footed ones (Chilopoda), by successive pairs of rings of the body uniting together. Fossil remains of the Chilopoda are first met with in the Jura period.

The third and last class of the Arthropoda breathing through tracheæ, is that of the Flies, or Insects, in the narrow sense of the word (Insecta, or Hexapoda), the largest of all classes of animals, and next to that of Mammalia, also the most important. Although Flies develop a greater variety of genera and species than all other animals taken together, yet these are all in reality only superficial variations of a single type, which is entirely and constantly preserved in its essential characteristics. In all Flies the three divisions of the trunk—head, breast (thorax), and hinder body—are quite distinct. The hinder body, or abdomen, as in the case of spiders, has no articulated appendages. The central division, the breast or thorax, has on its ventral side three pairs of legs, on its back two pairs of wings. It is true that, in very many Flies, one or both pairs of wings have become reduced in size or have even entirely disappeared; but the comparative anatomy of Flies distinctly shows that this deficiency has arisen only gradually by the degeneration of the wings, and that all the Flies existing at present are derived from a common, primary Fly, which possessed three pairs of legs and two pairs of wings. (Compare p. 256.) These wings, which so strikingly distinguish Flies from all other Arthropoda, probably arose, as has been already shown, out of the tracheate gills which may still be observed in the larvæ of the ephemeral flies (Ephemera) which live in water.

The head of Flies universally possesses, besides the eyes, a pair of articulated feelers, or antennæ, and also three jaws upon each side of the mouth. These three pairs of jaws, although they have arisen in all Flies from the same original basis, by different kinds of adaptation, have become changed to very varied and remarkable forms in the various orders, and are therefore employed for distinguishing and characterizing the main divisions of the class. In the first place, we may distinguish two main divisions, namely, Flies with chewing mandibles (Masticantia) and Flies with sucking mouths (Sugentia). On a closer examination each of these two divisions may again be divided into two sub-groups. Among chewing Flies, or Masticantia, we may distinguish the biting and the licking ones. Biting flies (Mordentia) comprise the most ancient and primæval winged Flies, the gauzy-winged (Neuroptera), straight-winged (Orthoptera), and beetles (Coleoptera). Licking flies (Lambentia) are represented by the one order of skin-winged (Hymenoptera) Flies. We distinguish two groups of Sucking Flies, or Sugentia, namely, those which prick and those which sip. There are two orders of pricking Flies (Pungentia), those with half wings (Hemiptera) and gnats and blow-flies (Diptera); butterflies are the only sipping Flies (Sorbentia), Lepidoptera.

Biting Flies, and indeed the order of Primæval Flies (Archiptera, or Pseudoneuroptera) are nearest akin to the still living Flies, and include the most ancient of all Flies, the primary forms of the whole class (hence also those of all Tracheata). Among them are, first of all, the Ephemeral Flies (Ephemera) whose larvæ which live in water, in all probability still show us in their trachea-gills the organs out of which the wings of Flies were originally developed. This order further contains the well known dragon-flies, or Libellula, the wine-glass sugar mites (Lepisma), the hopping Flies with bladder-like feet (Physopoda), and the dreaded Termites, fossil remains of which are found even in coal. The order of Gauze-winged Flies (Neuroptera), probably developed directly out of the primæval Flies, which differ from them only by their perfect series of transformations. Among them are the gauze-flies (Planipennia), caddis-flies (Phryganida), and fan-flies (Strepsiptera). Fossil Flies, which form the transition from the primæval Flies (Libellula) to the gauze-winged (Sialidæ), are found even in coal (Dictyophylebia).

The order of Straight-winged Flies (Orthoptera) developed at an early period out of another branch of the primæval Flies by differentiation of the two pairs of wings. This division is composed of one group with a great variety of forms—cockroaches, grasshoppers, crickets, etc. (Ulonata)—and of a smaller group consisting only of the well-known earwigs (Labidura), which are characterised by nippers at the hinder end of their bodies. Fossil remains of cockroaches, as well as of crickets and grasshoppers, have been found in coal.

Fossil remains of the fourth order of Biting Flies, beetles (Coleoptera) likewise occur in coal. This extremely comprehensive order—the favourite one of amateurs and collectors—shows more clearly than any other what infinite variety of forms can be developed externally by adaptation to different conditions of life, without the internal structure and the original form of the body being in any way essentially changed. Beetles have probably developed out of a branch of the straight-winged Flies, from which they differ only in their transformations (larva, pupa, etc.).

The one order of Licking Flies, namely, the interesting group of the Bees, or Skin-winged Flies (Hymenoptera), is closely allied to the four orders of biting Flies. Among them are those Flies which have risen to such an astonishing degree of mental development, of intellectual perfection, and strength of character, by their extensive division of labour, formation of communities and states, and surpass in this not merely most invertebrate animals, but even most animals in general. This may be said especially of all ants and bees, also of wasps, leaf-wasps, wood-wasps, gall-wasps, etc. They are first met with in a fossil state in the oolites, but they do not appear in greater numbers until the tertiary period. Probably these insects developed either out of a branch of the primæval Flies or the gauze-winged Flies.

Of the two orders of Pricking Flies (Hemiptera and Diptera), that containing the Half-winged Flies (Hemiptera), also called Beaked Flies (Rhynchota), is the older of the two. It includes three sub-orders, viz., the leaf-lice (Homoptera), the bugs (Heteroptera), and lice (Pediculina). Fossil remains of the first two classes are found in the oolites; but an ancient Fly (Eugereon) is found in the Permian system, and seems to indicate the derivation of the Hemiptera from the Neuroptera. Probably the most ancient of the three sub-orders of the Hemiptera are the Homoptera, among which, besides the actual leaf-lice, are the shield-lice, leaf-fleas, and leaf-crickets, or Cicadæ. Lice have probably developed out of two different branches of Homoptera, by continued degeneration (especially by the loss of wings); bugs, on the other hand, by the perfecting and differentiation of the two pairs of wings.

The second order of pricking flies, namely, the Two-winged Flies (Diptera), are also found in a fossil state in the oolites, together with Half-winged Flies; but they probably developed out of the Hemiptera by the degeneration of the hind wings. In Diptera the fore wings alone have remained perfect. The principal portion of this order consists of the elongated gnats (Nemocera) and of the compact blow-flies and house-flies (Brachycera), the former of which are probably the older of the two. However, remains of both are found in the oolitic period. The two small groups of lice-flies (Pupipara) forming chrysales, and the hopping-fleas (Aphaniptera), probably developed out of the Diptera by degeneration resulting from parasitism.

The eighth and last order of Flies, and at the same time the only one with mouth-parts adapted to sipping liquids, consists of moths and butterflies (Lepidoptera). This order appears, in several morphological respects, to be the most perfect class of Flies, and accordingly was the last to develop. For we only know of fossil remains of this order from the tertiary period, whereas the three preceding orders extend back to the oolites, and the four biting orders even to the coal period. The close relationship between some moths (Tineæ) and (Noctuæ), and some caddis-flies (Phryganida) renders it probable that butterflies have developed from this group, that is, out of the order of Gauze-winged Flies, or Neuroptera.

The whole history of Flies, and, moreover, the history of the whole tribe of Arthropoda, essentially confirms the great laws of differentiation and perfecting which, according to Darwin's theory of selection, must be considered as the necessary results of Natural Selection. The whole tribe, so rich in forms, begins in the Archilithic period with the class of Crabs breathing by gills, and with the lowest Primæval Crabs, or Archicaridæ. The form of these Primæval Crabs, which were developed out of segmented worms, is still approximately preserved by the remarkable Nauplius, in the common larval stage of so many Crabs. Out of the Nauplius, at a later period, the curious Zoëa was developed, which is the common larval form of all the higher or mailed crabs (Malacostraca), and, at the same time, possibly of that Arthopod which at first breathed through tracheæ, and became the common ancestor of all Tracheata. This Devonian ancestor, which must have originated between the end of the Silurian and the beginning of the Coal period, was probably most closely related to the still living Primæval Flies, or Archiptera. Out of these there developed, as the main tribe of the Tracheata, the class of Flies, from the lowest stage of which the spiders and centipedes separated as two diverging branches. Throughout a long period there existed only the four biting orders of Flies—the Primæval flies, Gauze-wings, Straight-wings, and the Beetles, the first of which is probably the common primary form of the three others. It was only at a much later period that the Licking, Pricking, and Sipping flies developed out of the Biting ones, which retained the original form of the three pairs of jaws most distinctly. The following table will show once more how these orders succeeded one another in the history of the earth.

PEDIGREE AND HISTORY OF THE ANIMAL KINGDOM

III. Vertebrate Animals.

Not one of the natural groups of organisms—which, we have designated as tribes, or phyla, on account of the blood-relationship of all the species included in them—is of such great and exceeding importance as the tribe of Vertebrate Animals. For, according to the unanimous opinion of all zoologists, man also is a member of the tribe; and his whole organization and development cannot possibly be distinguished from that of other Vertebrate animals. But as from the individual history of human development, we have already recognized the undeniable fact that, in developing out of the egg, man at first does not differ from other Vertebrate animals, and especially from Mammals, we must necessarily come to the conclusion, in regard to the palæontological history of his development, that man has, historically, actually developed out of the lower Vertebrata, and that he is directly derived from lower mammals. This circumstance, together with the many high interests which, in other respects, entitle the Vertebrata to more consideration than other organisms, justifies us in examining the pedigree of the Vertebrata and its expression in the natural system, with special care.

Fortunately, the records of creation, which must in all cases be our guide in establishing pedigrees, are especially complete in this important animal tribe, from which our own race has arisen. Even at the beginning of our century Cuvier's comparative anatomy and palæontology, and Bär's ontogeny of the Vertebrate animals, had brought us to a high level of accurate knowledge on this matter. Since then it is especially due to Johannes Müller's and Rathke's investigations in comparative anatomy, and

105

most recently to those of Gegenbaur and Huxley, that our knowledge of the natural relationships among the different groups of Vertebrata has become enlarged. It is especially Gegenbaur's classical works, penetrated as they are throughout with the fundamental principles of the Theory of Descent, which have demonstrated that the material of comparative anatomy receives its true importance and value only by the application of the Theory of Descent, and this in the case of all animals, but especially in that in the Vertebrate tribe. Here, as everywhere else, analogies must be traced to Adaptation, homologies to Transmission by Inheritance. When we see that the limbs of the most different Vertebrata, in spite of their exceedingly different external forms, nevertheless possess essentially the same internal structure; when we see that in the arm of a man and ape, in the wing of a man or a bird, in the breast fins of whales and sea-dragons, in the fore-legs of hoofed animals and frogs, the same bones always lie in the same characteristic position, articulation and connection—we can only explain this wonderful agreement and homology by the supposition of a common transmission by inheritance from a single primary form. On the other hand, the striking differences of these homologous bodily parts proceed from adaptation to different conditions of existence. (Compare Plate IV.)

Ontogeny, or the individual history of development, like comparative anatomy, is of especial importance to the pedigree of the Vertebrata. The first stages of development arising out of the egg are essentially identical in all Vertebrate animals, and retain their agreement the longer, the nearer the respective Vertebrate animal forms, when fully developed, stand to one another in the natural system, that is, in the pedigree. How far this agreement of germ forms, or embryos, extends, even in the most highly developed Vertebrate animals, I have already had occasion to explain (vol i. pp. 306-309). The complete agreement in form and structure, for example, in the embryos of a man and a dog, of a bird and a tortoise, existing in the stages of development represented on Plates II. and III., is a fact of incalculable importance, and furnishes us with the most important data for the construction of their pedigree.

Finally, the palæontological records of creation are also of especial value in the case of these same Vertebrate animals; for their fossil remains belong for the most part to the bony skeleton, a system of organs which is of the utmost importance for understanding their general organization. It is true that here, as in all other cases, the fossil records are exceedingly imperfect and incomplete, but more important remains of extinct Vertebrate animals have been preserved in a fossil state, than of most other groups of animals; and single fragments frequently furnish the most important hints as to the relationship and the historical succession of the groups.

The name of Vertebrate Animals (Vertebrata), as I have already said, originated with the great Lamarck, who towards the end of the last century comprised under this name, Linnæus' four higher classes of animals, viz. Mammals, Birds, Amphibious animals, and Fishes, Linnæus' two lower classes, Insects and Worms, Lamarck contrasted to the Vertebrata as Invertebrata, later also called Evertebrata.

The division of the Vertebrata into the four classes above named was retained also by Cuvier and his followers, and in consequence by many zoologists down to the present day. But in 1822 Blanville, the distinguished anatomist, found out by comparative anatomy—which Bär did almost at the same time from the ontogeny of Vertebrata—that Linnæus' class of Amphibious animals was an unnatural union of two very different classes. These two classes were separated as early as 1820, by Merrin, as two main groups of Amphibious animals, under the names of Pholidota and Batrachia. The Batrachia, which are at present (in a restricted sense) called Amphibious animals, comprise Frogs, Salamanders, gilled Salamanders, Cæcilia, and the extinct Labyrinthodonta. Their entire organization is closely allied to that of Fishes. The Pholidota, or Reptiles, on the other hand, are much more closely allied to Birds. They comprise lizards, serpents, crocodiles, and tortoises, and the groups of the mesolithic Dragons, Flying reptiles, etc.

In conformity with this natural division of Amphibious animals into two classes, the whole tribe of Vertebrate animals was divided into two main groups. The first main group, containing Amphibious animals and Fishes, breathe throughout their lives, or in early life, by means of gills, and are therefore called gilled Vertebrata (Branchiata, or Anallantoida). The second main group—Reptiles, Birds, and Mammals—breathe at no period of their lives through gills, but exclusively through lungs, and hence may appropriately be called Gill-less, or Vertebrata with lungs (Abranchiata, or Allantoida). However correct this distinction may be, still we cannot remain satisfied with it if we wish to arrive at a true natural system of the vertebrate tribe, and at a right understanding of its pedigree. In this case, as I have shown in my General Morphology, we are obliged to distinguish three other classes of Vertebrate animals, by dividing what has hitherto been regarded as the class of fishes into four distinct classes. (Gen. Morph. vol. ii. Plate VII. pp. 116-160.)

The first and lowest of these classes comprises the Skull-less animals (Acrania), or animals with tubular hearts (Leptocardia), of which only one representative now exists, namely, the remarkable little Lancelet (Amphioxus lanceolatus). Nearly allied to this is the second class, that of the Single-nostriled animals (Monorrhina), or Round-mouthed animals (Cyclostoma), which includes the Hags (Myxinoida) and Lampreys (Petromyzonta). The third class contains only the genuine Fish (Pisces): the Mud-fishes (Dipneusta) are added to these as a fourth class, and form the transition from Fish to Amphibious animals. This distinction, which, as will be seen immediately, is very important for the genealogy of the Vertebrate animals, increases the original number of Vertebrate classes from four to eight.

In most recent times a ninth class of Vertebrata has been added to these eight classes. Gegenbaur's recently published investigations in comparative anatomy prove that the remarkable class of Sea-dragons (Halisauria), which have hitherto been included among Reptiles, must be considered quite distinct from these, and as a separate class which branched off from the Vertebrate stock, even before the Amphibious animals. To it belong the celebrated large Ichthyosauri and Plesiosauri of the oolitic and chalk periods, and the older Simosauri of the Trias period, all of which are more closely allied to Fish than to Amphibious animals.

These nine classes of Vertebrate animals are, however, by no means of the same genealogical value. Hence we must divide them, as I have already shown in the Systematic Survey on p. 133, into four distinct main-classes or tribes. In the first place, the three highest classes, Mammals, Birds, and Reptiles, may be comprised as a natural main-class under the name of Amnion animals (Amnionata). The Amnion-less animals (Anamnionata), naturally opposed to them as a second main-class, include the four classes of Batrachians, Sea-dragons, Mud-fish, and Fishes. The seven classes just named, the Anamnionata as well as the Amnionata, agree among one another in numerous characteristics, which distinguish them from the two lowest classes (the single-nostriled and tubular-hearted animals). Hence we may unite them in the natural main group of Double-nostriled animals (Amphirrhina). Finally, these Amphirrhina on the whole are much more closely related to those animals with round mouths or single nostrils than to the skull-less or tube-hearted animals. We may, therefore, with full justice class the single and double-nostriled animals into one principal main group, and contrast them as animals with skulls (Craniota), or bulbular hearts (Pachycardia), to the one class of skull-less animals, or animals with tubular hearts. This classification of the Vertebrate animals proposed by me renders it possible to obtain a clear survey of the nine classes in their most important genealogical relations. The systematic relationship of these groups to one another may be briefly expressed by the following table.

A.Skull-less Animals
(Acrania)
1. Tubular hearts
1. Leptocardia
B.Animals with
Skulls (Craniota) or Thick Hearts
(Pachycardia)
a. Single nostriled animals
Monorrhina

2. Round-mouths 2. Cyclostoma
b. Double nostriled animals
Amphirrhina
I. Non-Amnionate
Anamnia
3. Fish 3. Pisces
4. Mud-fish 4. Dipneusta
5. Sea-dragons 5. Halisauria
6. Batrachians 6. Amphibia
II. Amnionate
Amniota
7. Reptiles 7. Reptilia
8. Birds 8. Aves
9. Mammals 9. Mammalia

The only one representative of the first class, the small lanceolate fish, or Lancelet (Amphioxus lanceolatus) (Plate XIII. Fig. B), stands at the lowest stage of organization of all the Vertebrate animals known to us. This exceedingly interesting and important animal, which throws a surprising light upon the older roots of our pedigree, is evidently the last of the Mohicans—the last surviving representative of a lower class of Vertebrate animals, very rich in forms, and very highly developed during the primordial period, but which unfortunately could leave no fossil remains on account of the absence of all solid skeleton. The Lancelet still lives widely distributed in different seas; for instance, in the Baltic, North Sea, and Mediterranean, where it generally lies buried in the sand on flat shores. The body, as the name indicates, has the form of a narrow lanceolate leaf, pointed at both extremities. When full grown it is about two inches long, of a white colour and semi-transparent. Externally, the little lanceolate animal is so little like a vertebrate animal that Pallas, who first discovered it, regarded it as an imperfect naked snail. It has no legs, and neither head, skull, nor brain. Externally, the fore end of the body can be distinguished from the hinder end only by the open mouth. But still the Amphioxus in its internal structure possesses those most important features, which distinguish all Vertebrate animals from all Invertebrate animals, namely, the spinal rod and spinal marrow. The spinal rod (Chorda dorsalis) is a straight, cylindrical, cartilaginous staff, pointed at both ends, forming the central axis of the internal skeleton, and the basis of the vertebral column. Directly above the spinal rod, on its dorsal side, lies the spinal marrow (medulla spinalis), likewise originally a straight but internally hollow cord, pointed at both ends. This forms the principal piece and centre of the nervous system in all Vertebrate animals. (Compare above 200 vol. i. p. 303.) In all Vertebrate animals without exception, man included, these important parts of the body during the embryological development out of the egg, originally begin in the same simple form, which is retained throughout life by the Amphioxus. It is only at a later period that the brain develops by the expansion of the fore end of the spinal marrow, and out of the spinal rod the skull which encloses the brain. As these two important organs do not develop at all in the Amphioxus, we may justly call the class represented by it, Skull-less animals (Acrania), in opposition to all the others, namely, to the animals with skulls (Craniota). The Skull-less animals are generally called tubular-hearted (Leptocardia), because a centralized heart does not as yet exist, and the blood is circulated in the body by the contractions of the tubular blood-vessels themselves. The Skulled animals, which possess a centralized, thick-walled, bulb-shaped heart, ought then by way of contrast to be called bulbular-hearted animals (Pachycardia).

Animals with skulls and central hearts evidently developed gradually in the later primordial period out of those without skulls and with tubular hearts. Of this the ontogeny of skulled animals leaves no doubt. But whence are these same skull-less animals derived? It is only very lately that an exceedingly surprising answer has been given to this important question. From Kowalewsky's investigations, published in 1867, on the individual development of the Amphioxus and the adhering Sea-squirts (Ascidia) belonging to the class of mantled animals (Tunicata), it has been proved that the ontogenies of these two entirely different looking animal-forms agree in the first stage of development in a most remarkable manner. The freely swimming larvæ of the Ascidians (Plate XII. Fig. A) develop the undeniable beginning of a spinal marrow (Fig. 5 g) and of a spinal rod (Fig. 5 c), and this moreover in entirely the same way as does the Amphioxus. (Plate XIII. Fig. B.) It is true that in the Ascidians these most important organs of the Vertebrate animal-body do not afterwards develop further. The Ascidians take on a retrograde transformation, become attached to the bottom of the sea, and develop into shapeless lumps, which when looked upon externally would scarcely be supposed to be animals. (Plate XIII. Fig. A.) But the spinal marrow, as the beginning of the central nervous system, and the spinal rod, as the first basis of the vertebral column, are such important organs, so exclusively characteristic of Vertebrate animals, that we may from them with certitude infer the true blood relationship of Vertebrate with Tunicate animals. Of course we do not mean to say by this, that Vertebrate animals are derived from Tunicate animals, but merely that both groups have arisen out of a common root, and that the Tunicates, of all the Invertebrata, are the nearest blood relations of the Vertebrates. It is quite evident that genuine Vertebrate animals developed progressively during the primordial period (and the skull-less animals first) out of a group of worms, from which the degenerate Tunicate animals arose in another and a retrograde direction. (Compare the more detailed explanation of Plates XII. and XIII. in the Appendix.)

Out of the Skull-less animals there developed, in the first instance, a second low class of Vertebrate animals, which still stands far below that of fish, and which is now represented only by the Hags (Myxinoida) and Lampreys (Petromyzonta). This class also, on account of the absence of all solid parts, could, unfortunately, as little as the Skull-less animals leave fossil remains. From its whole organization and ontogeny it is quite evident that it represents a very important intermediate stage between the Skull-less animals and Fishes, and that its few still existing members are only the last surviving remains of a probably very highly developed animal group which existed towards the end of the primordial period. On account of the curious mouth possessed by the Hags and Lampreys, which they use for sucking, the whole class is usually called Round-mouthed animals (Cyclostoma). The name of Single-nostriled animals (Monorrhina) is still more characteristic. For all Cyclostoma possess a simple, single nasal tube, whereas, in all other Vertebrate animals (with the exception of the Amphioxus) the nose consists of two lateral halves, a right and a left nostril. We are therefore enabled to comprise these latter (Anamnionata and Amnionata) under the heading, double-nostriled animals (Amphirrhina). All the Amphirrhina possess a fully developed jaw-skeleton (upper and under jaw), whereas it is completely wanting in the Monorrhina.

Apart also from the peculiar nasal formation, and the absence of jaws, the Single-nostriled animals are distinguished from those with double nostrils by many peculiarities. Thus they want the important sympathetic nervous system, and the spleen which the Amphirrhina possess. Of the swimming bladder, and the two pairs of legs—which all double-nostriled animals have, at least in their embryonic conditions—not a trace exists in the Single-nostriled animals, which is the case also in the Skull-less animals. Hence, we are surely justified in completely separating the Monorrhina, as we have separated the Skull-less animals, from the Fishes, with which they have hitherto been erroneously classed.

We owe our first accurate knowledge of the Monorrhina, or Cyclostoma, to the great zoologist, Johannes Müller of Berlin; his classical work on the "Comparative Anatomy of the Myxinoida" forms the foundation of our modern views on the structure of the Vertebrate animals. He distinguished two distinct groups among the Cyclostoma, which we shall consider as sub-classes.

The first sub-class consists of the Hags (Hyperotreta, or Myxinoida). They live in the sea as parasites upon other fish, into whose skin they penetrate (Myxine, Bdellostoma). Their organ of hearing has only one annular canal, and their single nasal tube penetrates the palate. The second sub-class, that of Lampreys, or Prides (Hyperoartia, or Petromyzontia) is more highly developed. It includes the well-known Lamperns, or Nine-eyes, of our rivers (Petromyzon fluviatilis), with which most persons are acquainted.

They are represented in the sea by the frequently larger marine or genuine Lampreys (Petromyzon marinus). The nasal tube of these single-nostriled animals does not penetrate the palate, and in the auricular organ there are two annular canals.

All existing Vertebrate animals, with the exception of the Monorrhina and Amphioxus just mentioned, belong to the group which we designate as Double-nostriled animals (Amphirrhina). All these animals possess (in spite of the great variety in the rest of their forms) a nose consisting of two lateral halves, a jaw-skeleton, a sympathetic nervous system, three annular canals connected with the auricular sac, and a spleen. Further, all Double-nostriled animals possess a bladder-shaped expansion of the gullet, which, in Fish, has developed into the swimming bladder, but in all other Double-nostriled animals into lungs. Finally, in all Double-nostriled animals there exist in the youngest stage of growth the beginnings of two pairs of extremities, or limbs, a pair of fore legs, or breast fins, and a pair of hinder legs, or ventral fins. One of these pairs of legs sometimes degenerates (as in the case of eels, whales, etc.), or both pairs of legs (as in Cæciliæ and serpents) either degenerate or entirely disappear; but even in these cases there exists some trace of their original beginning in an early embryonic period, or the useless remains of them may be found in the form of rudimentary organs. (Compare above, vol. i. p. 13.)

From all these important indications we may conclude with full assurance that all double-nostriled animals are derived from a single common primary form, which developed either directly or indirectly during the primordial period out of the Monorrhina. This primary form must have possessed the organs above mentioned, and also the beginning of a swimming bladder and of two pairs of legs or fins. It is evident, that of all still living double-nostriled animals, the lowest forms of sharks are most closely allied to this long since extinct, unknown, and hypothetical primary form, which we may call the Primary Double-nostriled animals (Proselachii). We may therefore look upon the group of primæval fish, or Selachii, to which the Proselachii probably belonged, as a primary group, not only of the Fish class, but of the whole main-class of double-nostriled animals.

The class of Fish (Pisces) with which we accordingly begin the series of Double-nostriled animals, is distinguished from the other six classes of the series by the swimming bladder never developing into lungs, but acting only as a hydrostatic apparatus. Agreeing with this, we find that in fish the nose is formed by two blind holes in front of the mouth, which never pierce the palate so as to open into the cavity of the mouth. In the other six classes of double-nostriled animals, both nostrils are changed into air passages which pierce the palate, and thus conduct air to the lungs. Genuine fish (after the exclusion of the Dipneusta) are accordingly the only double-nostriled animals which exclusively breathe through gills and never through lungs. In accordance with this, they all live in water, and both pairs of their legs have retained the original form of paddling fins.

Genuine fish are divided into three distinct sub-classes, namely, Primæval fish, Ganoid fish, and Osseous fish. The oldest of these, where the original form has been most faithfully preserved, is that of the Primæval fish (Selachii). Of these there still exist Sharks (Squali), and Rays (Rajæ), which are classed together as cross-mouthed fishes (Plagiostomi), and the strange and grotesquely formed Sea-cats, or Chimæracei (Holocephali). These primary fish of the present day, which are met with in all seas, are only poor remains of the prevailing animal groups, rich in forms, which the Selachii formed in the earlier periods of the earth's history, and especially during the palæolithic period. Unfortunately all Primæval fish possess a cartilaginous, never a completely osseous skeleton, which is but little, if at all, capable of being petrified. The only hard parts of the body which could be preserved in a fossil state, are the teeth and fin-spikes. These are found in the older formations in such quantities, varieties, and sizes, that we may, with certainty, infer a very considerable development of Primæval fish in those remote ages. They are even found in the Silurian strata, which contain but few remains of other Vertebrata, such as Enamelled fish (and these only in the most recent part, that is, in the upper Silurian). By far the most important and interesting of the three orders of Primæval fish are Sharks; of all still living double-nostriled animals, they are probably most closely allied to the original primary form of the whole group, namely, to the Proselachii. Out of these Proselachii, which probably differed but little from genuine Sharks, Enamelled fish, and the present Primæval fish, in all probability, developed in one direction, and the Dipneusta, Sea-dragons, and Amphibia in another.

The Ganoid, or Enamelled fish (Ganoides), in regard to their anatomy stand midway between the Primæval and the Osseous fish. In many characteristics they agree with the former, and in many others with the latter. Hence, we infer that genealogically they form the transition from Primæval to Osseous fish. The Ganoids are for the most part extinct, and more nearly so than the Primæval fish, whereas they were developed in great force during the entire palæolithic and mesolithic periods. Ganoid fish are divided into three legions according to the form of their external covering, namely, Mailed, Angular-scaled, and Round-scaled. The Mailed Ganoid fish (Tabuliferi) are the oldest, and are directly allied to the Selachii, out of which they originated. Fossil remains of them, though rare, are found even in the upper Silurian (Pteraspis ludensis of the Ludlow strata). Gigantic species of them, coated with strong bony plates, are found in the Devonian system. But of this legion there now lives only the small order of Sturgeons (Sturiones), including the Spade-sturgeons (Spatularidæ), and those Sturgeons (Accipenseridæ) to which belong, among others, the Huso, which yields isinglass, or sturgeon's sound, and the Caviar-sturgeon, whose eggs we eat in the shape of caviar, etc. Out of the mailed Ganoid fish, the angular and round-scaled ones probably developed as two diverging branches. The Angular-scaled Ganoid fish (Rhombiferi)—which can be distinguished at first sight from all other fish by their square or rhombic scales—are at present represented only by a few survivors, namely, the Finny Pike (Polypterus) in African rivers (especially the Nile), and by the Bony Pike (Lepidosteus) in American rivers. Yet during the palæolithic and the first half of the mesolithic epochs this legion formed the most numerous group of fishes. The third legion, that of Round-scaled Ganoid fish (Cycliferi), was no less rich in forms, and lived principally during the Devonian and Coal periods. This legion, of which the Bald Pike (Amia), in North American rivers, is the only survivor, was especially important, inasmuch as the third sub-class of fish, namely, Osseous fish, developed out of it.

Osseous fish (Teleostei) include the greater portion of the fish of the present day. Among these are by far the greater portion of marine fish, and all of our fresh-water fish except the Ganoid fish just mentioned. This class is distinctly proved by numerous fossils to have arisen about the middle of the Mesolithic epoch out of Ganoid fish, and moreover out of the Round-scaled, or Cycliferi. The Thrissopidæ of the Oolitic period (Thrissops, Leptolepis, Tharsis), which are most closely allied to the herrings of the present day, are probably the oldest of all Osseous fish, and have directly arisen out of Round-scaled Ganoid fish, closely allied to the existing Amia. In the older Osseous fish of the legion called Physostomi, as also in the Ganoides, the swimming bladder throughout life was connected with the throat by a permanent air passage (a kind of windpipe). This is still the case with all the fish belonging to this legion, namely, with herrings, salmon, carp, shad, eels, etc. However, during the chalk period this air passage, in some of the Physostomi, became constricted and closed, and the swimming bladder was thus completely separated from the throat. Hence there arose a second legion of Osseous fish, the Physoclisti, which did not attain their actual development until the tertiary epoch, and soon far surpassed the Physostomi in variety. To this legion belong most of the sea fish of the present day, especially the large families of the Turbot, Tunny, Wrasse, Crowfish, etc., further, the Lock-jaws (Plectognathi), Trunk fish, and Globe-fish and the Bushy-gills (Lophobranchi), viz., Pipe-fish, and Sea-horses. There are, however, only very few Physoclisti among our river fish, for instance, Perch and Sticklebacks; the majority of river fish are Physostomi.

Midway between genuine Fish and Amphibia is the remarkable class of Mud-fish, or Scaly Sirens (Dipneusta, or Protopteri). There now exist only a few representatives of this class, namely, the American Mud-fish

(Lepidosiren paradoxa) in the region of the river Amazon, and the African Mud-fish (Protopterus annectens) in different parts of Africa. A third large Salamander-fish (Ceratodus Fosteri) has lately been discovered in Australia. During the dry season, that is in summer, these strange animals bury themselves in a nest of leaves in the dry mud, and then breathe air through lungs like the Amphibia. But during the wet season, in winter, they live in rivers and bogs, and breathe water through gills like fish. Externally, they resemble fish of the eel kind, and are like them covered with scales; in many other characteristics also—in their internal structure, their skeleton, extremities, etc.—they resemble Fish more than Amphibia. But in certain features they resemble the Amphibia, especially in the formation of their lungs, nose, and heart. There is consequently an endless dispute among zoologists, as to whether the Mud-fish are genuine Fish or Amphibia. Distinguished zoologists have expressed themselves in favour of both opinions. But in fact, owing to the complete blending of characteristics which they present, they belong neither to the one nor to the other class, and are probably most correctly dealt with as a special class of Vertebrata, forming the transition between Fishes and Amphibians. The still living Dipneusta are probably the last surviving remains of a group which was formerly rich in forms, but has left no fossil traces on account of the want of a solid skeleton. In this respect, these animals are exactly like the Monorrhina and the Leptocardia. However, teeth are found in the Trias which resemble those of the living Ceratodus. Possibly the extinct Dipneusta of the palæolithic period, which developed in the Devonian epoch out of primæval fish, must be looked upon as the primary forms of the Amphibia, and thus also of all higher Vertebrata. At all events the unknown forms of transition—from Primæval fish to Amphibia—were probably very like the Dipneusta.

A very peculiar class of Vertebrate animals, long since extinct, and which appears to have lived only during the secondary epoch, is formed by the remarkable Sea-dragons (Halisauria, or Enaliosauria, also called Nexipoda, or Swimming-footed animals). These formidable animals of prey inhabited the mesolithic oceans in great numbers, and were of most peculiar forms, sometimes from thirty to forty feet in length. From many and excellently preserved fossil remains and impressions, both of the entire body of Sea-dragons as well as of single parts, we have become very accurately acquainted with the structure of their bodies. They are usually classed among Reptiles, whilst some anatomists have placed them in a much lower rank, as directly allied to Fish. Gegenbaur's recently published investigations, which place the structure of their limbs in a true light, have led to the surprising conclusion that the Sea-dragons form quite an isolated group, differing widely both from Reptiles and Amphibia as well as from Fish. The skeleton of their four legs, which are transformed into short, broad, paddling fins (like those of fish and whales) furnishes us with a clear proof that the Halisauria branched off from the main-stock of Vertebrata at an earlier period than the Amphibia. For Amphibia, as well as the three higher classes of Vertebrata, are all derived from a common primary form, which possessed only five toes or fingers on each leg. But the Sea-dragons have (either distinctly developed or in a rudimentary condition as parts of the skeleton of the foot) more than five fingers, as have also the Selachians or Primæval fish. On the other hand, they breathed air through lungs, like the Dipneusta, although they always swam about in the sea. They, therefore, perhaps, in conjunction with the Dipneusta, branched off from the Selachii, but did not develop into higher Vertebrata; they form an extinct lateral line of the pedigree, which has died out.

The more accurately known Sea-dragons are classed into three orders, distinct enough one from the other, namely, Primæval Dragons, Fish Dragons, and Serpent Dragons. The Primæval Dragons (Simosauria) are the oldest Sea-dragons, and lived only during the Trias period. The skeletons of many different genera of them are met with in the German limestone known as "Muschel-kalk." They seem upon the whole to have been very like the Plesiosauria, and are, consequently, sometimes united with them into one order as Sauropterygia. The Serpent Dragons (Plesiosauria) lived in the oolitic and chalk periods together with the Ichthyosauria. They were characterised by an uncommonly long thin neck, which was frequently longer than the whole body, and carried a small head with a short snout. When their arched neck was raised they must have looked very like a swan; but in place of wings and legs they had two pairs of short, flat, oval-paddling fins.

The body of the Fish Dragons (Ichthyosauria) was of an entirely different form; these animals may be opposed to the two preceding orders under the name of Fish-finners (Ichthyopterygia). They possessed a very long extended body, like a fish, and a heavy head with an elongated, flat snout, but a very short neck. Externally, they were probably very like porpoises. Their tail was very long, whereas it was very short in the members of the preceding orders. Also both pairs of paddling fins are broader and show very different structure from that seen in the other two orders. Probably the Fish Dragons and Serpent Dragons developed as two diverging branches out of the Primæval Dragons; but it is also possible that the Plesiosauria alone originated out of the Simosauria, and that the Ichthyosauria were lower off-shoots from the common stock. At all events, they must all be directly, or indirectly derived from the Selachii, or Primæval fish.

The succeeding classes of Vertebrata, the Amphibia and the Amniota (Reptiles, Birds, and Mammals), owing to the characteristic structure which they all exhibit of five toes to each foot, may all be derived from a common primary form, which originated from the Selachii, and which possessed five toes on each of its four limbs. When we find a less number of toes than five, we can show that the missing ones must have been lost in the course of time by adaptation. The oldest known Vertebrata with five toes are the Batrachias (Amphibia). We divide this class into two sub-classes, namely, mailed Batrachians and naked Batrachians, the first of which is distinguished by the body being covered with bony plates or scales.

The first and elder sub-class of Amphibia consists of the Mailed Batrachians (Phractamphibia), the oldest land living Vertebrata of which fossil remains exist. Well-preserved fossil remains of them occur in the coal, especially of those with Enamelled heads (Ganocephala), which are most closely allied to fish, namely, the Archegosaurus of Saarbruck, and the Dendrerpeton of North America. There then follow at a later period the gigantic Labyrinth-toothed animals (Labyrinthodonta), which are represented in the Permian system by Zygosaurus, but at a later period, more especially in the Trias, by Mastodonsaurus, Trematosaurus, Capitosaurus, etc. The shape of these formidable rapacious animals seems to have been between that of crocodiles, salamanders, and frogs, but in their internal structure they were more closely related to the two latter, while by their solid coat of mail, formed of strong bony plates, they resembled the first animals. These gigantic mailed Batrachians seem to have become extinct towards the end of the Triassic period. No fossil remains of mailed Batrachia are known during the whole of the subsequent periods. However, the still living blind Snakes, or Cæciliæ (Peromela)—small-scaled Phractamphibia of the form and the same mode of life as the earth-worm— prove that this sub-class continued to exist, and never became completely extinct.

The second sub-class of Amphibia, the naked Batrachia (Lissamphibia), probably originated even during the primary and secondary epochs, although fossil remains of them are first found in the tertiary epoch. They are distinguished from mailed Batrachia by possessing a naked smooth, and slimy skin, entirely without scales or coat of mail. They probably developed either out of a branch of the Phractamphibia, or out of the same common root with them. The ontogeny of the three still living orders of naked Batrachia—the gilled Batrachia, tailed Batrachia, and frog Batrachia— distinctly repeats the historical course of development of the whole sub-class. The oldest forms are the gilled Batrachia (Sozobranchia), which retain throughout life the original primary form of naked Batrachia, and possess a long tail, together with water-breathing gills. They are most closely allied to the Dipneusta, from which, however, they differ externally by the absence of the coat of scales. Most gilled Batrachia live in North America: among others of the class is the Axolotl, or Siredon, already mentioned. (Compare above, vol. i. p. 241.) In Europe the order is only represented by one form, the celebrated "Olm" (Proteus anguinus), which inhabits the grotto of

Adelsberg and other caves in Carinthia, and which, from living in the dark, has acquired rudimentary eyes which can no longer see (vol. i. p. 13). The order of Tailed Batrachia (Sozura) have developed out of the gilled Batrachia by the loss of external gills; the order includes our black and yellow spotted land Salamander (Salamandra maculata), and our nimble aquatic Salamanders (Tritons). Many of them—for instance, the celebrated giant Salamanders in Japan (Cryptobranchus Japonicus)—still retain the gill-slits, although the gills themselves have disappeared. All of them, however, retain the tail throughout life. Tritons occasionally—when forced to remain in water always—retain their gills, and thus remain at the same stage of development as gilled Batrachia. (Compare above, vol. i. p. 241.) The third order, the tailless or frog-like Batrachia (Anura), during their metamorphosis, not only lose their gills, with which in early life (as so-called tadpoles) they breathe in water, but also the tail with which they swim about. During their ontogeny, therefore, they pass through the course of development of the whole sub-class, they being at first Gilled Batrachia, then Tailed Batrachia, and finally Frog-like Batrachia. The inference from this is evidently, that Frog-like Batrachia developed at a later period out of Tailed Batrachia, as the latter had developed out of Gilled Batrachia which originally existed alone.

In passing from the Amphibia to the next class of Vertebrata, namely, Reptiles, we observe a very considerable advance in the progress of organization. All the double-nostriled animals (Amphirrhina) up to this time considered, and more especially the two larger classes of Fish and Batrachia, agree in a number of important characteristics, which essentially distinguish them from the three remaining classes of Vertebrata—Reptiles, Birds, and Mammals. During the embryological development of these latter, a peculiarly delicate covering, the first fœtal membrane, or amnion, which commences at the navel, is formed round the embryo; this membrane is filled with the amnion-water, and encloses the embryo or germ in the form of a bladder. On account of this very important and characteristic formation, we may comprise the three most highly developed classes of Vertebrata under the term Amnion-animals (Amniota). The four classes of double-nostriled animals which we have just considered, in which the amnion is wanting (as is the case in all lower Vertebrate animals, single-nostriled and skull-less animals), may on the other hand be opposed to the others as amnion-less animals (Anamnia).

The formation of the fœtal membrane, or amnion, which distinguishes reptiles, birds, and mammals from all other Vertebrata, is evidently a very important process in their ontogeny, and in the phylogeny which corresponds with it. It coincides with a series of other processes, which essentially determine the higher development of Amnionate animals. The first of these important processes is the total loss of gills, for which reason the Amniota, under the name of Gill-less animals (Ebranchiata), were formerly opposed to all other Vertebrate animals which breathed through gills (Branchiata). In all the Vertebrate already discussed, we found that they either always breathed through gills, or at least did so in early life, as in the case of Frogs and Salamanders. On the other hand, we never meet with a Reptile, Bird, or Mammal which at any period of its existence breathes through gills, and the gill-arches and openings which do exist in the embryos, are, during the course of the ontogeny, changed into entirely different structures, viz., into parts of the jaw-apparatus and the organ of hearing. (Compare above, vol. i. p. 307.) All Amnionate animals have a so-called cochlea in the organ of hearing, and a "round window" corresponding with it. These parts are wanting in the Amnion-less animals; moreover, their skull lies in a straight line with the axis of the vertebral column. In Amniotic animals the base of the skull appears bent in on the abdominal side, so that the head sinks upon the breast. (Plate III. Fig. C, D, G, H.) The organs of tears at the side of the eye also first develop in the Amniota.

The question now is, When did this important advance take place in the course of the organic history of the earth? When did the common ancestor of all Amniota develop out of a branch of the Non-amniota, to wit, out of the branch of the Amphibia?

To this question, the fossil remains of Vertebrata do not give us a very definite, but still they do give an approximate, answer. For with the exception of two lizard-like animals found in the Permian system (the Proterosaurus and Rhopalodon), all the fossil remains of Amniota, as yet known, belong to the secondary, tertiary, and quaternary epochs. With regard to the two Vertebrata just named, it is still doubtful whether they are genuine reptiles, or perhaps Amphibia of the salamander kind. Their skeleton alone is known to us, and even this not perfectly. Now as we know nothing of the characteristic features of their soft parts, it is quite possible that the Proterosaurus and Rhopalodon were non-amnionate animals more closely allied to Amphibia than to Reptiles; possibly they belonged to the transition form between the two classes. But, on the other hand, as undoubted fossil remains of Amniota have been found as early as the Trias, it is probable that the main class of Amniota first developed in the Trias, that is, in the beginning of the Mesolithic epoch. As we have already seen, this very period is evidently one of the most important turning points in the organic history of the earth. The palæolithic fern forests were then replaced by the pine forests of the Trias period; important transformations then took place in many of the classes of Invertebrata. Articulated marine lilies (Colocrina) developed out of the plated ones (Phatnocrina.) The Autechinidæ, or sea-urchins with only twenty rows of plates, took the place of the palæolithic Palechinidæ, the sea-urchins with more than twenty rows of plates. The Cystideæ, Blastoideæ, Trilobita, and other characteristic groups of Invertebrata of the primary period became extinct. It is no wonder that transforming conditions of adaptation powerfully influenced the Vertebrate tribes also in the beginning of the Trias period, and caused the origin of Amniotic animals.

If, however, the two Lizard and Salamander-like animals of the Permian system, the Proterosaurus and Rhopalodon, are considered genuine Reptiles, and consequently the most ancient Amniota, then the origin of this main class must necessarily have taken place in the preceding period, towards the end of the primary, namely, in the Permian period. However, all other remains of Reptiles, which were formerly believed to have been found in the Permian and the Coal system, or even in the Devonian system, have been proved to be either not remains of Reptiles at all, or to belong to a more recent date (for the most part to the Trias). (Compare Plate XIV.)

The common hypothetical primary form of all Amniotic animals, which we may call Protamnion, and which was possibly nearly related to the Proterosaurus, very probably stood upon the whole mid-way between salamanders and lizards, in regard to its bodily formation. Its descendants divided at an early period into two different lines, one of which became the common primary form of Reptiles and Birds, the other the primary form of Mammals.

Of all the three classes of Amniota, Reptiles (Reptilia, or Pholidota, also called Sauria in the widest sense), remain at the lowest stage of development, and differ least from their ancestors, the Amphibia. Hence they were formerly universally included among them, although their whole organization is much more like that of Birds than Amphibia. There now exist only four orders of Reptiles, namely,—Lizards, Serpents, Crocodiles, and Tortoises. They, however, form but a poor remnant of the exceedingly various and highly developed host of Reptiles which lived during the Mesolithic, or Secondary epoch, and predominated over all other Vertebrata. The immense development of Reptiles during the Secondary epoch is so characteristic that we could as well name it after those animals as after the Gymnosperms (p. 111). Twelve of the twenty-seven sub-orders, given on the accompanying table, and four of the eight orders, belong exclusively to the secondary period. These mesolithic groups are marked by an asterisk. All the orders, with the exception of Serpents, are found fossilized even in the Jura and Trias periods.

In the first order, that of Primary Reptiles, or Primary Creepers (Tocosauria), we class the extinct Thecodontia of the Trias, together with those Reptiles which we may look upon as the common primary form of the whole class. To the latter, which we may call Primæval Reptiles (Proreptilia), the Proterosaurus of the Permian system very probably belongs. The seven remaining orders must be considered as diverging branches, which have developed in different directions out of that common primary form. The Thecodontia of the Trias, the only positively known fossil forms of Tocosauria, were Lizards which seem to have been like the still living monitor lizards (Monitor, Varanus).

Of the four orders of reptiles now existing, and which, moreover, have alone represented the class since the beginning of the tertiary epoch, that of Lizards (Lacertilia) is probably most closely allied to the extinct Primary Reptiles, and especially through the monitors already named. The class of Serpents (Ophidia) developed out of a branch of the order of lizards, and this probably not until the beginning of the tertiary epoch. At least we at present only know of fossil remains of serpents from the tertiary strata. Crocodiles (Crocodilia) existed much earlier; the Teleosauria and Steneosauria belonging to the class are found fossil in large quantities even in the Jura; but the still living alligators are first met with in a fossil state in the chalk and tertiary strata. The most isolated of the four existing orders of reptiles consists of the remarkable group of Tortoises (Chelonia); fossils of these strange animals are first met with in the Jura. In some characteristics they are allied to Amphibia, in others, to Crocodiles, and by certain peculiarities even to Birds, so that their true position in the pedigree of Reptiles is probably far down at the root. The extraordinary resemblance of their embryos to Birds, manifested even at later stages of the ontogenesis, is exceedingly striking.

The four extinct orders of Reptiles show among one another, and, with the four existing orders just mentioned, such various and complicated relationships, that in the present state of our knowledge we are obliged to give up the attempt at establishing their pedigree. The most deviating and most curious forms are the Flying Reptiles (Pterosauria); flying lizards, in which the extremely elongated fifth finger of the hand served to support an enormous flying membrane. They probably flew about, in the secondary period, much in the same way as the bats of the present day. The smallest flying lizards were about the size of a sparrow; the largest, however, with a breadth of wing of more than sixteen feet, exceeded the largest of our living flying birds in stretch of wing (condor and albatross). Numerous fossil remains of them, of the long-tailed Rhamphorhynchia and of the short-tailed Pterodactylæ are found in all the strata of the Jura and Chalk periods, but in these only.

Not less remarkable and characteristic of the Mesolithic epoch was the group of Dragons (Dinosauria, or Pachypoda). These colossal reptiles, which attained a length of more than fifty feet, are the largest inhabitants of the land which have ever existed on our globe; they lived exclusively in the secondary epoch. Most of their remains are found in the lower cretaceous system, more especially in the Wealden formations of England. The majority of them were fearful beasts of prey (the Megalosaurus from twenty to thirty, the Pelorosaurus from forty to fifty feet in length). The Iguanodon, however, and some others lived on vegetable food, and probably played a part in the forests of the chalk period similar to that of the unwieldy but smaller elephants, hippopotami, and rhinoceroses of the present day.

The Beaked Reptiles (Anomodontia), likewise also long since extinct, but of which very many remarkable remains are found in the Trias and Jura, were perhaps closely related to the Dragons. Their jaws, like those of most Flying Reptiles and Tortoises, had become changed into a beak, which either possessed only degenerated rudimentary teeth, or no teeth at all. In this order, if not in the preceding one, we must look for the primary parents of the bird class, which we may call Bird Reptiles (Tocornithes). Probably very closely related to them was the curious, kangaroo-like Compsognathus from the Jura, which in very important characteristics already shows an approximation to the structure of birds.

The class of Birds (Aves), as already remarked, is so closely allied to Reptiles in internal structure and by embryonal development, that they undoubtedly originated out of a branch of this class. Even a glance at Plates II. and III. will show that the embryos of birds at a time when they already essentially differ from the embryos of Mammals, are still scarcely distinguishable from those of Tortoises and other Reptiles. The cleavage of the yolk is partial in the case of Birds and Reptiles, in Mammals it is total. The red blood-cells of the former possess a kernel, those of the latter do not. The hair of Mammals develops in closed follicles in the skin, but the feathers of birds and also the scales of reptiles develop in hillocks on the skin. The lower jaw of the latter is much more complicated than that of Mammals; the latter do not possess the quadrate bone of the former. Whereas in Mammals (as in the case of Amphibia) the connection between the skull and the first neck vertebra is formed by two knobbed joints, or condyles, in Birds and Reptiles those have become united into a single condyle. The two last classes may therefore justly be united into one group as Monocondylia, and contrasted to Mammals, or Dicondylia.

The deviation of Birds from Reptiles, in any case, first took place in the mesolithic epoch, and this moreover probably during the Trias. The oldest fossil remains of birds are found in the upper Jura (Archæopteryx). But there existed, even in the Trias period, different Saurians (Anomodonta) which in many respects seem to form the transition from the Tocosauria to the primary ancestors of Birds, the hypothetical Tocornithes. Probably these Tocornithes were scarcely distinguishable from other beaked lizards in the system, and were closely related to the kangaroo-like Compsognathus from the Jura of Solenhofen. Huxley classes the latter with the Dinosauria, and believes them to be the nearest relations to the Tocornithes.

The great majority of Birds—in spite of all the variety in the colouring of their beautiful feathery dress, and in the formation of their beaks and feet—are of an exceedingly uniform organization, in much the same way as are the class of insects. The bird form has adapted itself on all sides to the external conditions of existence, without having thereby in any way essentially deviated from the strict hereditary type of its characteristic structure. There are only two small groups, the feather-tailed birds (Saururæ) and those of the ostrich kind, which differ considerably from the usual type of bird, namely, from those with keel-shaped breasts (Carinatæ), and hence the whole class may be divided into three sub-classes.

The first sub-class, the Reptile-tailed, or Feather-tailed Birds (Saururæ), are as yet known only through a single, and that an imperfect, fossil impression, which, however, in being the oldest and also a very peculiar fossil bird, is of great importance. This fossil is the Primæval Griffin, or Archæopteryx lithographica, of which as yet only one specimen has been found in the lithographic slate at Solenhofen, in the Upper Jura system of Bavaria. This remarkable bird seems on the whole to have been of the size and form of a large raven, especially as regards the legs, which are in a good state of preservation; head and breast unfortunately are wanting. The formation of the wings deviates somewhat from that of other birds, but that of the tail still more so. In all other birds the tail is very short and composed of but few short vertebræ; the last of these have grown together into a thin, bony plate standing perpendicularly, upon which the rudder-feathers of the tail are attached in the form of a fan. The Archæopteryx, however, has a long tail like a lizard, composed of numerous (20) long thin vertebræ, and on every vertebra are attached the strong rudder-feathers in twos, so that the whole tail appears regularly feathered. This same formation of the tail part of the vertebral column occurs transiently in the embryos of other birds, so that the tail of the Archæopteryx evidently represents the original form of bird-tail inherited from reptiles. Large numbers of similar birds with lizard-tails probably lived during the middle of the secondary period; accident has as yet, however, only revealed this one fossil.

The Fan-tailed, or Keel-breasted birds (Carinatæ), which form the second sub-class, comprise all living Birds of the present day, with the exception of those of the ostrich kind, or Ratitæ. They probably developed out of Feather-tailed Birds during the first half of the secondary period, namely, in the Jura or Chalk period, by the hinder tail vertebræ growing together, and by the tail becoming shortened. Only very few remains of them are known from the secondary period, and these moreover only out of the last section of it, namely, from the Chalk. These remains belong to a swimming bird of the albatross species, and a wading bird like a snipe. All the other fossil remains of birds as yet known have been found in the tertiary strata.

The Bushy-tailed, or Ostrich-like Birds (Ratitæ), also called Running Birds (Cursores), the third and last sub-class, is now represented only by a few living species, by the African ostrich with two toes, the American and Australian ostrich with three toes, by the Indian cassowary and the four-toed kiwi, or Apteryx, in New Zealand. The extinct giant birds of Madagascar (Æpyornis) and the New Zealand Dinornis, which were much larger than the still living ostriches, also belong to this group. The Birds of the ostrich kind—by giving up the habit of flying, by the degeneration of the muscles for flying resulting from this, and of the breast bone which serves as their support, and by the corresponding stronger development of the hinder legs for running—have probably arisen out of a branch of the Keel-breasted birds. But possibly, as Huxley thinks, they may be the nearest relations of the Dinosauria and of the Reptiles akin to them, especially of the Compsognathus; at all events, the common primary form of all Birds must be looked for among the extinct Reptiles.

PEDIGREE AND HISTORY OF THE ANIMAL KINGDOM

IV. Mammals.

There are only a few points in the classification of organisms upon which naturalists have always agreed. One of these few undisputed points is the privileged position of the class of Mammals at the head of the animal kingdom. The reason of this privilege consists partly in the special interest, also in the various uses and the many pleasures, which Mammals, more than all other animals, offer to man, and partly in the circumstance that man himself is a member of this class. For however differently in other respects man's position in nature and in the system of animals may have been regarded, yet no naturalist has ever doubted that man, at least from a purely morphological point of view, belongs to the class of Mammals. From this there directly follows the exceedingly important inference that man, by consanguinity also, is a member of this class of animals, and has historically developed out of long since extinct forms of Mammals. This circumstance alone justifies us here in turning our especial attention to the history and the pedigree of Mammals. Let us, therefore, for this purpose first examine the groups of this class of animals.

Older naturalists, especially considering the formation of the jaw and feet, divided the class of Mammals into a series of from eight to sixteen orders. The lowest stage of the series was occupied by the whales, which seemed to differ most from man, who stands at the highest stage, by their fish-like form of body. Thus Linnæus distinguished the following eight orders: (1) Cetæ (whales); (2) Belluæ (hippopotami and horses); (3) Pecora (ruminating animals); (4) Glires (gnawing animals and rhinoceroses); (5) Bestiæ (insectivora, marsupials, and various others); (6) Feræ (beasts of prey); (7)

Bruta (toothless animals and elephants); (8) Primates (bats, semi-apes, apes, and men). Cuvier's classification, which became the standard of most subsequent zoologists, did not rise much above that of Linnæus. Cuvier distinguished the following eight orders: (1) Cetacea (whales); (2) Ruminantia (ruminating animals); (3) Pachyderma (hoofed animals, with the exclusion of ruminating animals); (4) Edentata (animals poor in teeth); (5) Rodentia (gnawing animals); (6) Carnassia (marsupials, beasts of prey, insectivora, and bats); (7) Quadrumana (semi-apes and apes); (8) Bimana (man).

The most important advance in the classification of Mammals was made as early as 1816 by the eminent anatomist Blainville, who has already been mentioned, and who first clearly recognised the three natural main groups or sub-classes of Mammals, and distinguished them according to the formation of their generative organs as Ornithodelphia, Didelphia, and Monodelphia. As this division is now justly considered by all scientific zoologists to be the best, on account of solid foundation on the history of development, let us here keep to it also.

The first sub-class consists of the Cloacal Animals, or Breastless animals, also called Forked animals (Monotrema, or Ornithodelphia). This class is now represented only by two species of living mammals, both of which are confined to Australia and the neighbouring island of Van Diemen's land, namely, the well-known Water Duck-bill (Ornithorhynchus paradoxus) with the beak of a bird, and the less known Beaked Mole (Echidna hystrix), resembling a hedgehog. Both of these curious animals, which are classed in the order of Beaked Animals (Ornithostoma), are evidently the last surviving remnants of an animal group formerly rich in forms, which alone represented the Mammalia in the secondary epoch, and out of which the second sub-class, the Didelphia, developed later, probably in the Jurassic period. Unfortunately, we as yet do not know with certainty of any fossil remains of this most ancient primary group of Mammals, which we will call Primary Mammals (Promammalia). Yet they possibly comprise the oldest of all the fossil Mammalia known, namely, the Microlestes antiquus, of which animals, however, we as yet only know some few small molar teeth. These have been found in the uppermost strata of the Trias, in the Keuper, first in Germany (at Degerloch, near Stuttgart, in 1847), later also in England (at Frome), in 1858. Similar teeth have lately been found also in the North American Trias, and have been described as Dromatherium sylvestre. These remarkable teeth, from the characteristic form of which we can conclude that they belonged to an insectivorous mammal, are the only remains of mammals as yet found in the older secondary strata, namely, in the Trias. It is possible, however, that besides these many of the other mammalian teeth found in the Jura and Chalk systems, which are still generally ascribed to Marsupials, in reality belong to Cloacal Animals. This cannot be decided with certainty owing to the absence of the characteristic soft parts. In any case, numerous Monotrema, with well-developed teeth and cloaca, must have preceded the advent of Marsupial animals.

The designation, "Cloacal animals" (Monotrema), has been given to the Ornithodelphia on account of the cloaca which distinguishes them from all other Mammals; but which on the other hand makes them agree with Birds, Reptiles, and Amphibia, in fact, with the lower Vertebrata. The formation of the cloaca consists in the last portion of the intestinal canal receiving the mouth of the urogenital apparatus, that is, the united urinary and genital organs, whereas in all other Mammals (Didelphia as well Monodelphia) these organs have an opening distinct from that of the rectum. However, in these latter also the cloaca formation exists during the first period of their embryonal life, and the separation of the two openings takes place only at a later date (in man about the twelfth week of development). The Cloacal animals have also been called "Forked animals," because the collar-bones, by means of the breast bone, have become united into one piece, similar to the well-known fork-bone, or merry-thought, in birds. In all other Mammals the two collar-bones remain separated in front and do not fuse with the breast bone. Moreover, the coracoid bones are much more strongly developed in the Cloacal animals than in the other Mammalia, and are connected with the breast bone.

In many other characteristics also—especially in the formation of their internal genital organs, their auricular labyrinth, and their brain—Beaked animals are more closely allied to the other Vertebrata than to Mammals, so that some naturalists have been inclined to separate them from the latter as a special class. However, like all other Mammals, they bring forth living young ones, which for a time are nourished with milk from the mother. But whereas in all other Mammals the milk issues through nipples, or teats, from the mammary glands, teats are completely wanting in beaked animals, and the milk comes simply out of a flat, sieve-like, perforated patch of the skin. Hence they may also be called Breastless or Teatless animals (Amasta). The curious formation of the beak in the two still living Beaked animals, which is connected with the suppression of the teeth, must evidently not be looked upon as an essential feature of the whole sub-class of Cloacal animals, but as an accidental character of adaptation distinguishing the last remnant of the class as much from the extinct main group, as the formation of a similar toothless snout distinguishes many toothless animals (for instance, the ant-eater) from the other placental animals. The unknown, extinct Primary Mammals, or Promammalia—which lived during the Trias period, and of which the two still living orders of Beaked animals represent but a single degenerate branch developed on one side—probably possessed a very highly developed jaw like the marsupial animals, which developed from them.

Marsupial, or Pouched Animals (Didelphia, or Marsupialia), the second of the three sub-classes of Mammals, form in every respect—both as regards their anatomy and embryology, as well as their genealogy and history—the transition between the other sub-classes—the Cloacal and Placental Animals. Numerous representatives of this group still exist, especially the well-known kangaroos, pouched rats, and pouched dogs; but on the whole this sub-class, like the preceding one, is evidently approaching its complete extinction, and the living members of the class are the last surviving remnants of a large group rich in forms, which represented the Mammalia during the more recent secondary and the earlier tertiary periods. The Marsupial Animals probably developed towards the middle of the Mesolithic epoch (during the Jura) out of a branch of the Cloacal Animals, and in the beginning of the Tertiary epoch again, the group of Placental Animals arose out of the Marsupials, and the latter then succumbed to the former in the struggle for life. All the fossil remains of Mammals known to us from the Secondary epoch, belong either exclusively to Marsupials, or partly perhaps to Cloacal animals. At that time Marsupials seem to have been distributed over the whole earth; even in Europe (France and England), well-preserved fossil remains of them have been found. On the other hand, the last off-shoots of the sub-class now living are confined to a very narrow tract of distribution, namely, to Australia, the Australasian, and a small part of the Asiatic Archipelago. There are also a few species still living in America, but at the present day not a single marsupial animal lives on the continent of Asia, Africa, or Europe.

The name of pouched animals is given to the class on account of the purse-shaped pouch (marsupium) existing in most instances on the abdominal side of the female animals, in which the mother carries about her young for a considerable time after their birth. This pouch is supported by two characteristic marsupial bones, also existing in Cloacal animals, but not in Placental animals. The young Marsupial animal is born in a much more imperfect form than the young Placental animal, and only attains the same degree of development which the latter possesses directly at its birth, after it has developed in the pouch for some time. In the case of the giant kangaroo, which attains the height of a man, the newly born young one, which has been carried in the maternal womb not much longer than five weeks, is not more than an inch in length, and only attains its essential development subsequently, in the pouch of the mother, where it remains about nine months attached to the nipple of the mammary gland.

The different divisions generally distinguished as families in the sub-class of Marsupial animals, deserve in reality the rank of independent orders, for they differ from one another in manifold differentiations of the jaw and limbs, in much the same manner, although not so sharply, as the various orders of Placental animals. In part they perfectly agree with the latter. It is evident that adaptation to similar conditions of life has effected entirely coincident or analogous transformations of the original fundamental form in the two sub-classes of Marsupials. According to this, about eight orders of Marsupial animals may be distinguished, the one half of the main group or legion of which are herbivorous, the other half carnivorous. The oldest fossil remains of the two legions (if the previously mentioned Microlestes and the Dromatherium are not included) occur in the Jurassic strata, namely, in the slates of Stonesfield, near Oxford. The slates belong to the Bath, or the Lower Oolite formation—strata which lie directly above the Lias, the oldest Jura formation. (Compare p. 15.) It is true that the remains of Marsupials found in the slates of Stonesfield, as well as those which were found later in the Purbeck strata, consist only of lower jaws. (Compare p. 29.) But fortunately the lower jaw is just one of the most characteristic parts of the skeleton of Marsupials. For it is distinguished by a hook-shaped process of the lower corner of the jaw turning downwards and backwards, which neither occurs in Placental nor in the (still living) Cloacal animals, and from the existence of this process on the lower jaws from Stonesfield, we may infer that they belonged to Marsupials.

Of Herbivorous marsupials (Botanophaga), only two fossils are as yet known from the Jura, namely, the Stereognathus ooliticus, from the slates of Stonesfield (Lower Oolite), and the Plagiaulax Becklesii, from the middle Purbeck strata (Upper Oolite). But in Australia there are gigantic fossil remains of extinct herbivorous Marsupials from the diluvial period (Diprotodon and Nototherium) which were far larger than the largest of the still living Marsupials. The Diprotodon Australis, whose skull alone is three feet long, exceeded even the river-horse, or Hippopotamus, in size and upon the whole resembled it in the unwieldy and clumsy form of body. This extinct group, which probably corresponded with the gigantic placental hoofed animals of the present day—the hippopotami and rhinoceroses—may be called Hoofed Marsupials (Barypoda). Closely allied to them is the order of kangaroos, or Leaping Marsupials (Macropoda), which all have seen in zoological gardens. In their shortened fore legs, their very lengthened hind legs, and very strong tail, which serves as a jumping pole, they correspond with the leaping mice in the class of Rodents. Their jaw, however, resembles that of horses, and their complex stomach that of Ruminants. A third order of Herbivorous Marsupials corresponds in its jaws to Rodents, and in its subterranean mode of life, especially, to digging mice. Hence they may be termed Rodent Marsupials, or root-eating pouched animals (Rhizophaga). They are now represented only by the Australian wombat (Phascolomys). A fourth and last order of Herbivorous Marsupials is formed by the climbing or Fruit-eating Marsupials (Carpophaga), whose mode of life and structure resembles partly that of squirrels, partly that of apes (Phalangista, Phascolarctus).

The second legion of Marsupials, the Carnivorous Marsupials (Zoophaga), is likewise divided into four main groups or orders. The most ancient of these is that of the primæval, or Insectivorous Marsupials (Cantharophaga). It probably includes the primary forms of the whole legion, and possibly also those of the whole sub-class. At least, all the lower jaws from Stonesfield (with the exception of the Stereognathus) belong to Insectivorous Marsupials, and the still living Myrmecobius is their nearest relative. But some of those oolitic Primæval Marsupials possessed a larger number of teeth than all the other known mammals, for each half of the lower jaw of the Thylacotherium contained sixteen teeth (three incisors, one canine tooth, six pseudo, and six genuine molars). If the upper jaw, which is unknown, had as many teeth, then the Thylacotherium had no less than sixty-four teeth, just double the number possessed by man. The Primæval Marsupials correspond, on the whole, with the Insectivora among Placental animals, which order includes hedgehogs, moles, and shrew-mice. A second order, which has probably developed out of a branch of the last, consists of the Snouted, or Toothless Marsupials (Edentula), which resemble the Toothless animals, or Edentata, among the Placental animals by their tube-shaped snout, their degenerated jaws, and their corresponding mode of life. On the other hand, the mode of life and formation of the jaws of Rapacious marsupials (Creophaga) correspond with those of the genuine Beasts of Prey, or Carnivora, among Placental animals. This order includes the pouched marten (Dasyurus) and the pouched wolf (Thylacinus) in Australia. Although the latter attains to the size of a wolf, it is but a dwarf in comparison with the extinct Australian pouched lions (Thylacoleo) which were at least as large as a lion, and possessed huge canine teeth more than two inches in length. Finally, the eighth and last order is formed by the marsupials with hands, or the Ape-footed Pouched animals (Pedimana), which live both in Australia and America. They are frequently kept in zoological gardens, especially the different species of the genus Didelphys, and are known by the name of pouched rats, bush rats, or opossums. The thumb on their hinder feet is opposable to the four other toes, as in a hand, and by this they are directly allied to the Semi-apes, or Prosimia, among Placental animals. It is possible that these latter are really next akin to the marsupials with hands, and that they have developed out of their long since extinct ancestors.

It is very difficult to discover the genealogy of Marsupials, and this more especially because we are but very imperfectly acquainted with the whole sub-class; and the Marsupials of the present day are evidently only the last remnants of a group that was at one time rich in forms. It is possible that Marsupials with hands, those with snouts, as well as rapacious Marsupials, developed as three diverging branches out of the common primary group of Primæval Marsupials. In a similar manner, on the other hand, the rodent, leaping, and hoofed Marsupials have perhaps arisen as three diverging branches out of the common herbivorous primary group, that is, out of the Climbing Marsupials. Climbing and Primæval Marsupials might, however, be two diverging branches of the common primary forms of all Marsupials, that is, of the Primary Marsupials (Prodidelphia), which originated during the older secondary period out of Cloacal animals.

The third and last sub-class of mammals comprises the Placental animals, or Placentals (Monodelphia, or Placentalia). It is by far the most important, comprehensive, and most perfect of the three sub-classes; for the class includes all the known mammalia, with the exception of Marsupials and Beaked animals. Man also belongs to this sub-class, and has developed out of its lower members.

Placental animals, as their name indicates, are distinguished from all other mammals, more especially by the formation of a so called placenta. This is a very peculiar and remarkable organ, which plays an exceedingly important part in nourishing the young one developing in the maternal body. The placenta (also called after-birth) is a soft, spongy, red body, which differs very much in form and size, but which consists for the most part of an intricate network of veins and blood vessels. Its importance lies in the exchange of substance between the nutritive blood of the maternal womb, or uterus, and the body of the germ, or embryo. (See vol. i. p. 298.) This very important organ is developed neither in marsupials nor in beaked animals. But placental animals are also distinguished from these two sub-classes by many other peculiarities, thus more especially by the absence of marsupial bones, by the higher development of the internal sexual organs, and by the more perfect development of the brain, especially of the so-called callous body or beam (corpus callosum), which, as the intermediate commissure, or transverse bridge, connects the two hemispheres of the large brain with each other. Placental animals also do not possess the peculiar hooked process of the lower jaw which characterizes Marsupials. The following classification (p. 246) of the most important characteristics of the three sub-classes will best explain how Marsupials, in these anatomical respects, stand midway between Cloacal and Placental animals.

Placental animals are more variously differentiated and perfected, and this, moreover, in a far higher degree, than Marsupials, and they have, on this account, long since been arranged into a number of orders, differing principally in the formation of the jaws and feet. But what is even of more importance than these, is the different development of the placenta, and the manner of its connection with the maternal uterus. For in the three lower orders of Placental animals, in Hoofed animals, Whales, and Toothless animals, the peculiar spongy membrane, which is called the deciduous membrane, or decidua, and which connects the maternal and the fœtal portions of the placenta, does not become developed. This takes place exclusively in the seven higher orders of Placental animals, and we may, therefore, according to Huxley, class them in the main group of Deciduata, or animals with decidua. They are contrasted with the three first-mentioned legions of indeciduous animals, or Indeciduata.

Three Sub-Classes of Mammals.	Cloacal Animals Monotrema or Ornithodelphia	Pouched Animals Marsupialia or Didelphia	Placental Animals Placentalia or Monodelphia
1. Cloaca formation	Constant	Embryonal	Embryonal
2. Nipples of the pectoral glands, or milk warts	Wanting	Existing	Existing
3. Fore collar bones, or clavicles, grown together in the middle, with the breast bone, and forming a forked bone	United	Not united	Not united
4. Marsupial bones	Existing	Existing	Wanting
5. Corpus callosum of the brain	Feebly developed	Feebly developed	Strongly developed
6. Placenta	Wanting	Wanting	Existing

But in the various orders of Placental animals the placenta differs not only in important internal differences of structure, which are connected with the absence or the presence of a decidua, but also in the external form of the placenta itself. In the Indeciduata it consists, in most cases, of numerous, single, scattered bunches or tufts of vessels, and hence this group may be called tufted placental animals (Villiplacentalia). In the Deciduata, however, the single tufts of vessels are united into a cake, which appears in two different forms. In the one case it surrounds the embryo in the form of a closed band or ring, so that only the two poles of the oval egg bladder are free of tufts; this is the case in animals of prey (Carnaria) and the pseudo-hoofed animals (Chelophora), which may consequently be comprised as girdled-placental animals (Zonoplacentalia). In the other Deciduata, to which man also belongs, the placenta is a simple round disc, and we therefore call them disc-placentals (Discoplacentalia). This group includes the five orders of Semi-apes, Gnawing animals, Insectivora, Bats, and Apes, from the latter of which, in the zoological system, man cannot be separated. It may be considered as quite certain, from reasons based upon their comparative anatomy and their history of development, that Placental animals first developed out of Marsupials, and that this very important development—the first origin of the placenta—probably took place in the beginning of the tertiary epoch, during the eocene period. But one of the most difficult questions in the genealogy of animals is the important consideration whether all Placental animals have arisen out of one or out of several distinct branches of Marsupials; in other words, whether the origin of the placenta occurred but once, or several times.

When, in my General Morphology, I for the first time endeavoured to establish the pedigree of Mammals, I here, as in most cases, preferred the monophyletic, or one-rooted, to the polyphyletic, or many-rooted, hypothesis of descent. I assumed that all Placental animals were derived from a single form of Marsupial animal, which, for the first time, began to form a placenta. In this case the Villiplacentals, Zonoplacentals, and Discoplacentals would perhaps have to be considered as three diverging branches of the common primary form of Placentals, or it might also be conceived that the two latter, the Deciduata, had developed only at a later period out of the Indeciduata, which on their part had arisen directly out of the Marsupials. However, there are also important reasons for the alternative; namely, that several groups of Placentals, differing from the beginning, arose out of several distinct groups of Marsupials, so that the placenta itself was formed several times independently. This opinion is maintained by Huxley, the most eminent English zoologist, and by many others. In this case the Indeciduata and the Deciduata would perhaps have to be considered as two completely distinct groups; then the order of Hoofed animals, as the primary group of the Indeciduata, might be supposed to have originated out of the Marsupial hoofed animals (Barypoda). Among the Deciduata, on the other hand, the order of Semi-apes, as the common primary form of the other orders, might possibly have arisen out of Handed Marsupials (Pedimana). But it is also conceivable that the Deciduata themselves have arisen out of several different orders of Marsupials, Animals of Prey out of Rapacious Marsupials, Gnawing animals out of Gnawing Marsupials, Semi-apes out of Handed Marsupials, etc. As we do not at present possess sufficient empiric material to solve this most difficult question, we must leave it and turn our attention to the history of the different orders of Placental animals, whose pedigree can often be very accurately established in detail.

We must, as already remarked, consider the order of Hoofed animals (Ungulata) as the primary group of the Indeciduata, or Tuft-placentals; the two other orders, Whales and Toothless animals, developed out of them, as two diverging groups, probably only at a later period, by adaptation to very different modes of life. But it is also possible that the animals poor in teeth (Edentata) may be of quite a different origin.

Hoofed animals are in many respects among the most important and the most interesting Mammals. They distinctly show that a true understanding of the natural relationship of animals can never be revealed to us merely by the study of living forms, but in all cases only by an equal consideration of their extinct and fossil blood-relations and ancestors. If, as is usually done, only the living Hoofed animals are taken into consideration, it seems quite natural to divide them into three entirely distinct orders, namely: (1) Horses, or Single-hoofed animals (Solidungula, or Equina); (2) Ruminating animals, or Double-hoofed (Bisulca, or Ruminantia); and (3) Thick-skinned, or Many-hoofed (Multungula, or Pachyderma). But as soon as the extinct Hoofed animals of the tertiary period are taken into consideration—of which animals we possess very numerous and important remains—it is seen that this division, but more especially the limitation of the Thick-skinned animals, is completely artificial, and that these three groups are merely top branches lopped from the pedigree of Hoofed animals, which are most closely connected by extinct intermediate forms. The one half of the Thick-skinned animals—rhinoceroses, tapirs, and palæotheria—manifest the closest relationships to horses, and have like them odd-toed feet; whereas the other half of the Thick-skinned animals—pigs, hippopotami, and anoplotheria—on account of their double-toed feet are much more closely allied to ruminating animals than to the former. Hence we must, in the first place, among Hoofed animals distinguish the two orders of Paired-hoofs and Odd-hoofs, as two natural groups, which developed as diverging branches out of the old tertiary primary group of Primary Hoofed animals, or Prochela.

The order of Odd-hoofed animals (Perissodactyla) comprises those Ungulata in which the middle (or third) toe of the foot is much more strongly developed than the others, so that it forms the actual centre of the hoof. This order includes the very ancient, common, primary group of all Hoofed animals, that is, the Primary-hoofed animals (Prochela), which are found in a fossil state in the oldest Eocene strata (Lophiodon, Coryphodon, Pliolophus). Directly allied to this group is that branch which is the actual primary form of the Odd-hoofed animals, namely, the Palæotheria, fossils of which occur in the upper Eocene and lower Miocene. Out of the Palæotheria, at a later period, the rhinoceroses (Nasicornia) and rhinoceros-horses (Elasmotherida) on the one hand, and the tapirs, lama-tapirs, and primæval horses, on the other, developed as two diverging branches. The long since extinct primæval horses, or Anchitheria, formed the transition from the Palæotheria and tapirs to the Miocene horses, or hipparions, which are closely allied to the genuine living horses.

The second main group of Hoofed animals, the order of Pair-hoofed animals (Artiodactyla), comprises those hoofed animals in which the middle (third) and fourth toe of the foot are almost equally developed, so that the space between the two forms the central line of the entire foot. The order is divided into two sub-orders—the Pig-shaped and the Cud-chewing, or Ruminating. The Pig-shaped (Chœromorpha) comprise in the first place the other branch of Primary-Hoofed-animals, the Anoplotheria, which we consider as the common primary form of all Pair-hoofed animals, or Artiodactyla (Dichobune, etc.). Out of the Anoplotheria arose, as two diverging branches, the primæval swine, or Anthracotheria, on the one hand, forming the transition to swine and river-horses, and the Xiphodonta on the other hand, forming the transition to Ruminating animals. The oldest Ruminating animals (Ruminantia) are the Primæval Stags, or Dremotheria, out of which, possibly, the stag-shaped (Elaphia), the hollow-horned (Cavicornia), and camels (Tylopoda), have developed as three diverging branches. Yet these latter are, in many respects, more allied to the Odd-hoofs than to the genuine Pair-hoofs. The accompanying systematic survey on p. 252, will show how the numerous families of Hoofed animals are grouped, in correspondence with this genealogical hypothesis.

It is probable that the remarkable legion of Whales (Cetacea) originated out of Hoofed animals, which accustomed themselves exclusively to an aquatic life, and thereby became transformed into the shape of fish. Although these animals seem externally very like many genuine Fish, yet they are, as even Aristotle perceived, genuine Mammals. By their whole internal structure—in so far as it has not become changed by adaptation to an aquatic life—they, of all known Mammals, are most closely allied to Hoofed animals, and more especially agree with them in the absence of the decidua and in the tufted placenta. Even at the present day the river-horse (Hippopotamus) constitutes a kind of transition form to the Sea Cows (Sirenia), and from this it seems most probable that the extinct primary forms of the Cetacea are most closely allied to the Sea Cows of the present day, and that they developed out of Pair-hoofed animals, which were related to the hippopotamus. Out of the order of Herbivorous whales (Phycoceta)—to which the sea cows belong, and which accordingly, very probably, contain the primary forms of the legion—the other order of Carnivorous whales (Sarcoceta) appears to have developed at a later period. But Huxley thinks that these latter were of quite a different origin, and that they arose out of the Carnaria through the Seals. Among the Sarcoceta, the extinct gigantic Zeuglodonta (Zeugloceta)—whose fossil skeletons some time ago excited great interest, it being thought that they were "sea serpents"—are probably only a peculiarly developed lateral branch of genuine whales (Autoceta), which comprise, besides the colossal whalebone whales, the cachalot or spermaceti whales, dolphins, narwhals, porpoises, etc.

The third legion of the Indeciduata, or Sparsi-placentalia, comprises the strange group of the animals poor in teeth (Edentata); it is composed of the two orders of burrowers and sloths. The order of Burrowers (Effodientia) consists of the two sub-orders of ant eaters (Vermilinguia), to which the scaled animals also belong, and the girdle animals (Cingulata), which were formerly represented by the gigantic Glyptodons. The order of Sloths (Tardigrada) consists of the two sub-orders of the small, still living dwarf sloths (Bradypoda), and of the extinct unwieldy giant sloths (Gravigrada). The enormous fossil remains of these colossal herbivora suggest that the whole legion is becoming extinct, and that the Edentata of the present day are but a poor remnant of the mighty order of the diluvial period. The close relations between the still living South American Edentata and the extinct gigantic forms which are found beside the latter on the same part of the globe, made such an impression upon Darwin on his first visit to South America, that they even then suggested to him the fundamental idea of the Theory of Descent. (See above, vol. i. p. 134.) But it is precisely the genealogy of this legion which is most difficult. The Edentata are perhaps nothing but a peculiarly developed lateral branch of the Ungulata; but it may also be that their root lies in quite another direction.

We now leave the first main group of Placental animals, the Indeciduata, and turn to the second main group, namely, the Deciduata, or animals with decidua, which are distinguished from the former by possessing a deciduous membrane, or decidua, during their embryonal life. We here meet with a very remarkable small group of animals, for the most part extinct, and which probably were the old tertiary (or eocene) ancestors of man. These are the Semi-apes, or Lemurs (Prosimiæ); these curious animals are probably the but little changed descendants of the primæval group of Placentalia which we have to consider as the common primary form of all Deciduata. They have hitherto been classed together in the same order with Apes which Blumenbach called Quadrumana (four-handed). However, I regard them as entirely distinct from these, not merely because they differ from all Apes, much more than do the most different Apes from one another, but also because they comprise most interesting transitional forms leading to the other orders of Deciduata. I conclude from this that the few still living Semi-apes, which moreover differ very much among one another, are the last surviving remnants of a primary group now almost extinct, but which was at one time rich in forms, and out of which all the other

Deciduata (possibly with the single exception of Beasts of Prey, and Pseudo-hoofed animals) have developed as diverging branches. The old primary group of Semi-apes has probably developed out of Handed or Ape-footed Marsupials (Pedimana), which are surprisingly like them in the transformation of their hinder feet into grasping hands. The primæval primary forms themselves (which probably originated in the eocene period) are of course long since extinct, as are also the greater portion of the transition-forms between them and all the other orders of Deciduata. However, individual remnants of the latter are preserved among the Semi-apes of the present day. Among these, the remarkable Finger-animal of Madagascar (Chiromys madagascariensis) constitutes the remnant of the group of the Leptodactyla and the transition to Rodents. The strange flying lemur in the South Sea and Sunda islands (Galeopithecus), the only remnant of the group of Pteropleura, forms a perfect intermediate stage between Semi-apes and Bats. The long-footed Semi-apes (Tarsius, Otolicnus) constitute the last remnant of that primary branch (Macrotarsi) out of which the Insectivora developed. The short-footed forms (Brachytarsi) are the medium of connection between them and genuine Apes. The Short-footed Semi-apes comprise the long-tailed Lemur, the short-tailed Lichanotus, and the Stenops, the latter of which seems to be very closely allied to the probable ancestors of man among the Semi-apes. The short-footed as well as the long-footed Prosimiæ live widely distributed over the islands of southern Asia and Africa, more especially in Madagascar; some live also on the continent of Africa. No Semi-ape, either living or in a fossil state, has as yet been found in America. They all lead a solitary, nocturnal kind of life, and climb about on trees. (Compare vol. i. p. 361.)

Among the six remaining orders of Deciduata, all of which are probably derived from long since extinct Semi-apes, the order of Gnawing animals (Rodentia), which is rich in forms, has remained at the lowest stage. Among these the squirrel-like animals (Sciuromorpha) stand nearest akin to the Pedimanous Marsupials. Out of this primary group the mouse-like animals (Myomorpha) and the porcupine-like animals (Hystricomorpha) developed probably as two diverging branches, the former of which are directly connected with the squirrel-like animals, by the eocene Myoxida, the latter by the eocene Psammoryctida. The fourth sub-order, the hare-like animals (Lagomorpha), probably developed only at a later period out of one of the other three sub-orders.

Very closely allied to the Rodentia is the remarkable order of Pseudo-hoofed animals (Chelophora). Of these there now live but two genera, indigenous to Asia and Africa, namely, Elephants (Elephas), and Rock Conies (Hyrax). Both have hitherto generally been classed among real Hoofed animals, or Ungulata, with which they agree in the formation of the feet. But an identical transformation of nails or claws into hoofs occurs also in genuine Rodentia and in certain hoofed Rodentia (Subungulata) which live exclusively in South America. Beside smaller forms (for example, guinea pigs and gold hares) the Subungulata also include the largest of all Rodentia, namely, the Capybara Rats, which are about four feet in length. The Rock Conies, which are externally very nearly akin to Rodents, especially to the hoofed Rodents, were formerly classed among Rodentia by some celebrated zoologists, as an especial sub-class (Lamnungia). Elephants, on the other hand, when not classed among Hoofed animals, were generally considered as the representatives of a special order which were called Trunked animals (Proboscidea). But the formation of the placentas of Elephants and of Hyrax agree in a remarkable manner, and are entirely distinct from those of Hoofed animals. These latter never possess a decidua, whereas Elephants and Hyrax are genuine Deciduata. Their placenta is indeed not of the form of a disc, but of a girdle, as in the case of Animals of Prey; it is very possible that the girdle-shaped placenta is but a secondary development of the discoplacenta. Thus, then, it might be thought that the Pseudo-hoofed animals have developed out of a branch of the Rodentia, and in a similar manner perhaps the Carnivora out of a branch of the Insectivora. At all events, Elephants and Hyrax in many respects, especially in the formation of important skeletal parts, of the limbs, etc., are more closely allied to the Rodentia, and more especially to hoofed Rodentia, than to genuine Hoofed animals. Moreover several extinct forms, especially the remarkable South American Arrow-toothed animals (Toxodontia), stand in many respects mid-way between Elephants and Rodentia. That the still living Elephants and Hyrax are but the last survivors of a group of Pseudo-hoofed animals, which was once rich in forms, is proved not only by the very numerous fossil species of Elephants and Mastodon (some of which are even larger, others also much smaller than the Elephants of the present day), but also by the remarkable miocene Dinotheria (Gonyognatha), between which and their next kindred, the Elephants, there must be a long series of unknown connecting intermediate forms. Taking all things into consideration, the most probable hypothesis which can be established at present as to the origin and the relationship of Elephants, Dinotheria, Toxodon, and Hyrax is, that they are the last survivors of a group of Pseudo-hoofed animals rich in forms, which developed out of the Rodentia, and probably out of relatives of the Subungulata.

The order of Insect Eaters (Insectivora) is a very ancient group, and is next akin to the common extinct primary form of the Deciduata, as well as to the Semi-apes of the present day. It has probably developed out of Semi-apes which were closely allied to the Long-footed Lemurs (Macrotarsi) of the present day. It is separated into two orders, Menotyphla and Lipotyphla; the Menotyphla are probably the older of the two, and are distinguished from the Lipotyphla by possessing an intestinal cœcum, or typhlon. The

Menotyphla include the climbing Tupajas of the Sunda Isles, and the leaping Macroscelides of Africa. The Lipotyphla are represented in our country by shrew mice, moles, and hedgehogs. The Insectivora, in the formation of their jaws and their mode of life, are nearly akin to Carnivora, but are, on the other hand, by their discoplacentas and by their large seminal vesicles, allied to Rodents.

It is probable that the order of Rapacious animals (Carnaria) developed out of a long since extinct branch of Insectivora, at the beginning of the Eocene period. It is a natural group, very rich in forms, but still of very uniform organization. The Rapacious animals are sometimes also called Girdle-placentals (Zonoplacentals), although the Pseudo-hoofed animals (Chelophora), in the same way, also deserve this designation. But as the latter, in other respects, are more closely allied to the Rodentia than to Carnaria, we have already discussed them in connection with the former. Animals of prey are divided into two, externally very different, but internally very closely related, sub-orders, namely, Land animals of prey and Marine animals of prey. The Land animals of prey (Carnivora) comprise bears, dogs, cats, etc., whose pedigree can be approximately guessed at by means of many extinct intermediate forms. The Marine animals of prey, or Seals (Pinnipedia), comprise sea bears, sea dogs, sea lions, and walruses. Although marine animals of prey appear externally very unlike land animals of prey, yet by their internal structure, their jaw and their peculiar girdle-shaped placenta, they are very nearly akin to them, and have evidently originated out of a branch of them, probably out of a kind of weasel (Mustelina). Even at the present day the fish otters (Lutra), and still more so the sea otters (Enhydris), present a direct form of transition to Seals, and clearly show how the bodies of land Carnivora are transformed into the shape of a Seal, by adaptation to an aquatic life, and how the steering fins of marine rapacious animals have arisen out of the legs of the former. The latter consequently stand in the same relation to the former as do the Whales to Hoofed animals among the Indeciduata. In the same way as the river-horse at present stands midway between the extreme branches of oxen and sea oxen, the sea otter still forms a surviving intermediate stage between the widely separated branches of dogs and sea dogs. In both cases the complete transformation of the external form, consequent upon adaptation to entirely different conditions of life, has not been able to efface the solid foundation of the inherited internal peculiarities.

According to Huxley's opinion, which has already been quoted, only the Herbivorous Whales (Sirenia) are derived from Hoofed animals; on the other hand, the Carnivorous Cetacea (Sarcoceta) are derived from the marine animals of prey; the Zeuglodonts would form a transition between the two latter. But in this case it would be difficult to understand the close anatomical relations which exist between the Herbivorous and Carnivorous

Cetacea. The strange peculiarities in the internal and external structure which so strikingly distinguish the two groups from all other mammals would then have to be regarded only as analogies (caused by the same kinds of adaptation), not as homologies (transmitted from a common primary form). The latter, however, strikes me as being by far the more probable, and hence I have left all the Cetacea among the Indeciduata as one group of kindred origin.

The remarkable order of Flying Mammals, or Bats (Chiroptera), stands near to the Carnaria as well as to the Insectivora. It has become strikingly transformed by adaptation to a flying mode of life, just as marine animals of prey have become modified by adaptation to a swimming mode of life. This order probably also originated out of the Semi-apes, with which it is even at present closely allied, through the flying lemurs (Galeopithecus). Of the two orders of flying animals, the insect-eating forms, or flying mice (Nycterides), probably developed out of those eating fruits, or flying foxes (Pterocynes); for the latter are, in many ways, more closely allied to Semi-apes than are the former.

We have now still to discuss the genuine Apes (Simiæ) as the last order of Mammals; but as, according to the zoological system, the human race belongs to this order, and as it undoubtedly developed historically out of a branch of this order, we shall devote a special chapter to a more careful examination of its pedigree and history.

ORIGIN AND PEDIGREE OF MAN

Of all the individual questions answered by the Theory of Descent, of all the special inferences drawn from it, there is none of such importance as the application of this doctrine to Man himself. As I remarked at the beginning of this treatise, the inexorable necessity of the strictest logic forces us to draw the special deductive conclusion from the general inductive law of the theory, that Man has developed gradually, and step by step, out of the lower Vertebrata, and more immediately out of Ape-like Mammals. That this doctrine is an inseparable part of the Theory of Descent, and hence also of the universal Theory of Development in general, is recognized by all thoughtful adherents of the theory, as well as by all its opponents who reason logically.

But if the doctrine be true, then the recognition of the animal origin and pedigree of the human race will necessarily affect more deeply than any other progress of the human mind the views we form of all human relations, and the aims of all human science. It must sooner or later produce a complete revolution in the conception entertained by man of the entire universe. I am firmly convinced that in future this immense advance in our knowledge will be regarded as the beginning of a new period of the development of Mankind. It can only be compared to the discovery made by Copernicus, who was the first who ventured distinctly to express the opinion, that it was not the sun which moved round the earth, but the earth round the sun. Just as the geocentric conception of the universe—namely, the false opinion that the earth was the centre of the universe, and that all its other portions revolved round the earth—was overthrown by the system of the universe established by Copernicus and his followers, so the anthropocentric conception of the universe—the vain delusion that Man is the centre of terrestrial nature, and that its whole aim is merely to serve him—is overthrown by the application (attempted long since by Lamarck) of the theory of descent to Man. As Copernicus' system of the universe was mechanically established by Newton's theory of gravitation, we see Lamarck's theory of descent attain its causal establishment by Darwin's theory of selection. This comparison, which is very interesting in many respects, I have discussed in detail elsewhere.

In order to carry out this extremely important application of the Theory of Descent to man, with the necessary impartiality and objectivity, I must above all beg the reader (at least for a short time) to lay aside all traditional and customary ideas on the "Creation of Man," and to divest himself of the deep-rooted prejudices concerning it, which are implanted in the mind in earliest youth. If he fail to do this, he cannot objectively estimate the weight of the scientific arguments which I shall bring forward in favour of the animal derivation of Man, that is, of his origin out of Ape-like Mammals. We cannot here do better than imagine ourselves with Huxley to be the inhabitants of another planet, who, taking the opportunity of a scientific journey through the universe, have arrived upon the earth and have there met with a peculiar two-legged mammal called Man, diffused over the whole earth in great numbers. In order to examine him zoologically, we should pack a number of the individuals of different ages and from different lands (as we should do with the other animals collected on the earth) into large vessels filled with spirits of wine, and on our return to our own planet we should commence the comparative anatomy of all these terrestrial animals quite objectively. As we should have no personal interest in Man, in a creature so entirely different from ourselves, we should examine and criticise him as impartially and objectively as we should the other terrestrial animals. In doing this we should, of course, in the first place refrain from all conjectures and speculations on the nature of his soul, or on the spiritual side of his nature, as it is usually called. We should occupy ourselves solely with his bodily structure, and with that natural conception of it which is offered by the history of his individual development.

It is evident that in order correctly to determine Man's position among the other terrestrial organisms we must, in the first place, follow the guidance of the natural system. We must endeavour to determine the position which belongs to Man in the natural system of animals as accurately and distinctly as possible. We shall then, if in fact the theory of descent be correct, be able from his position in the system to determine the real primary relationship, and the degree of consanguinity connecting Man with the animals most like him. The hypothetical pedigree of the human race will then follow naturally as the final result of this anatomical and systematic inquiry.

Now if, by means of comparative anatomy and ontogeny, we seek for man's position in that Natural System of animals which formed the subject of the last two chapters, the incontrovertible fact will at once present itself to us, that man belongs to the tribe, or phylum, of the Vertebrata. Every one of the characteristics, which so strikingly distinguish all the Vertebrata from all Invertebrata, is possessed by him. It has also never been doubted that of all the Vertebrata the Mammals are most closely allied to Man, and that he possesses all the characteristic features distinguishing them from all other Vertebrata. If then we further carefully examine the three different main groups or sub-classes of Mammals—the inter-connections of which were discussed in our last chapter—there cannot be the slightest doubt that Man belongs to the Placentals, and shares with all other Placentals, the important characteristics which distinguish them from Marsupials and from Cloacals. Finally, of the two main groups of placental Mammals, the Deciduata and the Indeciduata, the group of Deciduata doubtless includes Man. For the human embryo is developed with a genuine decidua, and is thus absolutely distinguished from all the Indeciduata. Among the Deciduata we distinguish two legions, the Zonoplacentalia, with girdle-shaped placenta (Beasts of Prey and Pseudo-hoofed animals), and the Discoplacentalia, with disc-shaped placenta (all the remaining Deciduata). Man possesses a disc-shaped placenta, like all Discoplacentalia; and thus our next question must be, What is man's position in this group?

In the last chapter we distinguished the following five orders of Discoplacentalia: (1) Semi-apes; (2) Rodents; (3) Insectivora; (4) Bats; (5) Apes. The last of these five orders, that of Apes, is, as every one knows, in every bodily feature far more closely allied to Man than the four others. Hence the only remaining question now is, whether, in the system of animals, Man is to be directly classed in the order of genuine Apes, or whether he is to be considered as the representative of a special sixth order of Discoplacentalia, allied to, but more advanced than, that of the Apes.

Linnæus in his system classed Man in the same order with genuine Apes, Semi-apes, and Bats, which he called Primates; that is, lords, as it were the highest dignitaries of the animal kingdom. But Blumenbach, of Göttingen, separated Man as a special order, under the name of Bimana, or two-handed, and contrasted him with the Apes and Semi-apes under the name of Quadrumana, or four-handed. This classification was also adopted by Cuvier and, consequently, by most subsequent zoologists. It was not until 1863 that Huxley, in his excellent work, the "Evidence as to Man's Place in Nature,"(26) showed that this classification was based upon erroneous ideas, and that the so-called "four-handed" Apes and Semi-apes are "two-handed" as much as man is himself. The difference between the foot and hand does not consist in the physiological peculiarity that the first digit or thumb is opposable to the four other digits or fingers in the hand, and is not so in the foot, for there are wild tribes of men who can oppose the first or large toe to the other four, just as if it were a thumb. They can therefore use their "grasping foot" as well as a so-called "hinder hand," like Apes. The Chinese boatmen row with this hinder hand, the Bengal workmen weave with it. The Negro, in whom the big toe is especially strong and freely moveable, when climbing seizes hold of the branches of the trees with it, just like the "four-handed" Apes. Nay, even the newly born children of the most highly developed races of men, during the first months of their life, grasp as easily with the "hinder hand" as with the "fore hand," and hold a spoon placed in its clutch as firmly with their big toe as with the thumb! On the other hand, among the higher Apes, especially the gorilla, hand and foot are differentiated as in man. (Compare Plate IV.)

The essential difference between hand and foot is therefore not physiological, but morphological, and is determined by the characteristic structure of the bony skeleton and of the muscles attached to it. The ankle-bones differ from the wrist-bones in arrangement, and the foot possesses three special muscles not existing in the hand (a short flexor muscle, a short extensor muscle, and a long fibular muscle). In all these respects, Apes and Semi-apes entirely agree with man, and hence it was quite erroneous to separate him from them as a special order on account of the stronger differentiation of his hand and foot. It is the same also with all the other structural features by means of which it was attempted to distinguish Man from Apes; for example, the relative length of the limbs, the structure of the skull, of the brain, etc. In all these respects, without exception, the differences between Man and the higher Apes are less than the corresponding differences between the higher and the lower Apes. Hence Huxley, for reasons based on the most careful and most accurate anatomical comparisons, arrives at the extremely important conclusion—"Thus, whatever system of organs be studied, the comparison of their modifications in the Ape series leads to one and the same result, that the structural differences which separate Man from the Gorilla and Chimpanzee are not so great as those which separate the Gorilla from the lower Apes." In accordance with this, Huxley, strictly following the demands of logic, classes Man, Apes, and Semi-apes in a single order, Primates, and divides it into the following seven families, which are of almost equal systematic value: (1) Anthropini (Man); (2) Catarrhini (genuine Apes of the Old World); (3) Platyrrhini (genuine American Apes); (4) Arctopitheci (American clawed Apes); (5) Lemurini (short-footed and long-footed Semi-apes, p. 255); (6) Chiromyini (p. 256); (7) Galeopithecini (Flying Lemurs, p. 256).

If we wish to arrive at a natural system, and consequently at the pedigree of the Primates, we must go a step further still, and entirely separate the Semi-apes, or Prosimiæ, (Huxley's last three families), from Genuine Apes, or Simiæ (the first four families). For, as I have already shown in my General Morphology, and explained in the last chapter, the Semi-apes differ in many and important respects from Genuine Apes, and in their individual forms are more closely allied to the various other orders of Discoplacentalia. Hence the Semi-apes must probably be considered as the remnants of the common primary group, out of which the other orders of Discoplacentalia, and, it may be, all Deciduata, have developed as two diverging branches. (Gen. Morph. ii. pp. 148 and 153.) But man cannot be separated from the order of Genuine Apes, or Simiæ, as he is in every respect more closely allied to the higher Genuine Apes than the latter are to the lower Genuine Apes.

Genuine Apes (Simiæ) are universally divided into two perfectly natural groups, namely, the Apes of the New World, or American Apes, and the Apes of the Old World, which are indigenous to Asia and Africa, and which formerly also existed in Europe. These two classes differ principally in the formation of the nose, and they have been named accordingly. American Apes have flat noses, so that the nostrils are in front, not below; hence they are called Flat Noses (Platyrrhini). On the other hand, the Apes of the Old World have a narrow cartilaginous bridge, and the nostrils turned downwards, as in man; they are, therefore, called Narrow Noses (Catarrhini). Further, the jaw, which plays an important part in the classification of Mammals, is essentially distinct in these two groups. All Catarrhinæ, or Apes of the Old World, have exactly the same jaws as Man, namely, in each jaw four incisors above and below, then on each side a canine tooth and five cheek teeth, of which two are pre-molars and three molars, altogether thirty-two teeth. But all Apes of the New World, all Platyrrhini, have four more cheek teeth, namely, three pre-molars and three molars on each side, above and below: they consequently possess thirty-six teeth. Only one small group forms an exception to this rule, namely, the Arctopitheci, or Clawed Apes, in whom the third molar has degenerated, and they accordingly have on each half of their jaw three pre-molars and two molars. They also differ from the other Platyrrhini by having claws on the fingers of their hands and the toes of their feet, not nails like Man and the other Apes. This small group of South American Apes, which includes among others the well-known pretty little Midas-monkey and the Jacchus, must probably be considered only as a peculiarly developed lateral branch of the Platyrrhini.

Now, if we ask what evidence can be drawn, as to the pedigree of Apes, from the above facts, we must conclude that all the Apes of the New World have developed out of one tribe, for they all possess the characteristic jaw and the nasal formation of the Platyrrhini. In like manner it follows that all the Apes of the Old World must be derived from one and the same common primary form, which possessed the same formation of nose and jaw as all the still living Catarrhini. Further, it can scarcely be doubted that the Apes of the New World, taken as an entire tribe, are either derived from those of the Old World, or (to express it more vaguely and cautiously) both are diverging branches of one and the same tribe of Apes. We also arrive at the exceedingly important conclusion—which is of the utmost significance in regard to Man's distribution on the earth's surface—that Man has developed out of the Catarrhini. For we cannot discover a zoological character distinguishing him in a higher degree from the allied Apes of the Old World than that in which the most divergent forms of this group are distinguished from one another. This is the important result of Huxley's careful anatomical examination of the question, and it cannot be too highly estimated. The anatomical differences between Man and the most human-like Catarrhini (Orang, Gorilla, Chimpanzee) are in every respect less than the anatomical differences between the latter and the lowest stages of Catarrhini, more especially the Dog-like Baboon. This exceedingly important conclusion is the result of an impartial anatomical comparison of the different forms of Catarrhini.

If, therefore, we recognise the natural system of animals as the guide to our speculations, and establish upon it our pedigree, we must necessarily come to the conclusion that the human race is a small branch of the group of Catarrhini, and has developed out of long since extinct Apes of this group in the Old World. Some adherents of the Theory of Descent have thought that the American races of Men have developed, independently of those of the Old World, out of American Apes. I consider this hypothesis to be quite erroneous, for the complete agreement of all mankind with the Catarrhini, in regard to the characteristic formation of the nose and jaws, distinctly proves that they are of the same origin, and that they developed out of a common root after the Platyrrhini, or American Apes, had already branched off from them. The primæval inhabitants of America, as is proved by numerous ethnographical facts, immigrated from Asia, and partly perhaps from Polynesia (or even from Europe).

There still exist great difficulties in establishing an accurate pedigree of the Human Race; this only can we further assert, that the nearest progenitors of man were tail-less Catarrhini (Lipocerca), resembling the still living Man-like Apes. These evidently developed at a late period out of tailed Catarrhini (Menocerca), the original form of Ape. Of those tail-less Catarrhini, which are now frequently called Man-like Apes, or Anthropoides, there still exist four different genera containing about a dozen different species.

The largest Man-like Ape is the famous Gorilla (called Gorilla engena, or Pongo gorilla), which is indigenous to the tropics of western Africa, and was first discovered by the missionary, Dr. Savage, in 1847, on the banks of the river Gaboon. Its nearest relative is the Chimpanzee (Engeco troglodytes, or Pongo troglodytes), also indigenous to western Africa, but considerably smaller than the Gorilla, which surpasses man in size and strength. The third of the three large Man-like Apes is the Orang, or Orang

Outang, indigenous to Borneo and the other Sunda Islands, of which two kindred species have recently been distinguished, namely, the large Orang (Satyrus orang, or Pithecus satyrus) and the small Orang (Satyrus morio, or Pithecus morio). Lastly, there still exists in southern Asia the genus Gibbon (Hylobates), of which from four to eight different species are distinguished. They are considerably smaller than the three first-named Anthropoides, and in most characteristics differ more from Man.

The tail-less Man-like Apes—especially since we have become more intimately acquainted with the Gorilla, and its connection with Man by the application of the Theory of Descent—have excited such universal interest, and called forth such a flood of writings, that there is no occasion for me here to enter into any detail about them. The reader will find their relations to Man fully discussed in the excellent works of Huxley,(26) Carl Vogt,(27) Büchner,(43) and Rolle.(28) I shall therefore confine myself to stating the most important general conclusion resulting from their thorough comparison with Man, namely, that each one of the four Man-like Apes stands nearer to Man in one or several respects than the rest, but that no one of them can in every respect be called absolutely the most like Man. The Orang stands nearest to Man in regard to the formation of the brain, the Chimpanzee in important characteristics in the formation of the skull, the Gorilla in the development of the feet and hands, and, lastly, the Gibbon in the formation of the thorax.

Thus, from a careful examination of the comparative anatomy of the Anthropoides, we obtain a similar result to that obtained by Weisbach, from a statistical classification and a thoughtful comparison of the very numerous and careful measurements which Scherzer and Schwarz made of the different races of Men during their voyage in the Austrian frigate Novara round the earth. Weisbach comprises the final result of his investigations in the following words: "The ape-like characteristics of Man are by no means concentrated in one or another race, but are distributed in particular parts of the body, among the different races, in such a manner that each is endowed with some heirloom of this relationship—one race more so, another less, and even we Europeans cannot claim to be entirely free from evidences of this relationship."5

I must here also point out, what in fact is self-evident, that not one of all the still living Apes, and consequently not one of the so-called Man-like Apes, can be the progenitor of the Human Race. This opinion, in fact, has never been maintained by thoughtful adherents of the Theory of Descent, but it has been assigned to them by their thoughtless opponents. The Ape-like progenitors of the Human Race are long since extinct. We may possibly still find their fossil bones in the tertiary rocks of southern Asia or Africa. In any case they will, in the zoological system, have to be classed in the group of tail-less Narrow-nosed Apes (Catarrhini Lipocerci, or

Anthropoides).

The genealogical hypotheses, to which we have thus far been led by the application of the Theory of Descent to Man, present themselves to every clearly and logically reasoning person as the direct results from the facts of comparative anatomy, ontogeny, and palæontology. Of course our phylogeny can indicate only in a very general way the outlines of the human pedigree. Phylogeny is the more in danger of becoming erroneous the more rigorously it is applied in detail to special animal forms known to us. However, we can, even now, with approximate certainty distinguish at least the following twenty-two stages of the ancestors of Man. Fourteen of these stages belong to the Vertebrata, and eight to the Invertebrate ancestors of Man (Prochordata.)

First Stage: Monera.

The most ancient ancestors of Man, as of all other organisms, were living creatures of the simplest kind imaginable, organisms without organs, like the still living Monera. They consisted of simple, homogeneous, structureless and formless little lumps of mucous or albuminous matter (protoplasm), like the still living Protamœba primitiva. (Compare vol. i. p. 186, Fig. 1.) The form value of these most ancient ancestors of man was not even equal to that of a cell, but merely that of a cytod (compare vol. i. p. 347); for, as in the case of all Monera, the little lump of protoplasm did not as yet possess a cell-kernel. The first of these Monera originated in the beginning of the Laurentian period by spontaneous generation, or archigony, out of so-called "inorganic combinations," namely, out of simple combinations of carbon, oxygen, hydrogen, and nitrogen. The assumption of this spontaneous generation, that is, of a mechanical origin of the first organisms from inorganic matter, has been proved in our thirteenth chapter to be a necessary hypothesis. (Compare vol. i. p. 338.) A direct proof of the earlier existence of this most ancient ancestral stage, based upon the fundamental law of biogeny, is possibly still furnished by the circumstance that, according to the assertions of many investigators, in the beginning of the development of the egg, the cell-kernel, or nucleus, disappears, and the egg-cell thus relapses to the lower stage of the cytod (Monerula, p. 124; relapse of the nucleated plastid into a non-nucleated condition). The assumption of this first stage is necessary for most important general reasons.

Second Stage: Amœbæ.

The second ancestral stage of Man, as of all the higher animals and plants, is formed by a simple cell, that is, a little piece of protoplasm enclosing a kernel. There still exist large numbers of similar "single-celled organisms." Among them the common, simple Amœbæ (vol. i. p. 188, Fig. 2) cannot have been essentially different from these progenitors. The form value of every Amœba is essentially the same as that still possessed by the egg of

Man, and by the egg of all other animals. (Vol. i. p. 189, Fig. 3.) The naked egg-cells of Sponges, which creep about exactly like Amœbæ, cannot be distinguished from them. The egg-cell of Man, which like that of most other animals is surrounded by a membrane, resembles an enclosed Amœba. The first single-celled animals of this kind arose out of Monera by the differentiation of the inner kernel and the external protoplasm; they lived in the earlier Primordial period. An irrefutable proof that such single-celled primæval animals really existed as the direct ancestors of Man, is furnished according to the fundamental law of biogeny (vol. i. p. 309) by the fact that the human egg is nothing more than a simple cell. (Compare p. 124.)

Third Stage: Synamœbæ.

In order to form an approximate conception of the organisation of those ancestors of Man which first developed out of the single-celled Primæval animals, it is necessary to trace the changes undergone by the human egg in the beginning of its individual development. It is just here that ontogeny guides us with the greatest certainty on to the track of phylogeny. We have already seen that the egg of Man (in the same way as that of all other Mammals), after fructification has taken place, falls by self-division into a mass of simple and equi-formal Amœba-like cells (vol. i. p. 190, Fig. 4 D). All these divided globules are at first exactly like one another, naked cells containing a kernel, but without covering; in many animals they show movements like those of the Amœbæ. This ontogenetic stage of development which we called Morula (p. 125), on account of its mulberry shape, is a certain proof that in the early primordial period there existed ancestors of man which possessed the form value of a mass of homogeneous, loosely connected cells. They may be called a community of Amœbæ (Synamœbæ). (Compare p. 127.) They originated out of the single-celled Primæval animals of the second stage by repeated self-division and by the permanent union of the products of this division.

Fourth Stage: Ciliated Larva (Planæada).

In the course of the ontogenesis of most of the lower animals, and also in that of the lowest Vertebrate animals, the Lanceolate Animals, or Amphioxus, there first develops out of the Morula (Frontispiece, Fig. 3) a ciliated larva (planula). Those cells, lying on the surface of the homogeneous mass of cells, extend hair-like processes, or fringes of hairs, which by striking against the water keep the whole body rotating. The round many-celled body thus becomes differentiated, in that the external cells covered with cilia differ from the non-ciliated internal cells (Frontispiece, Fig. 4). In Man and in all other Vertebrate animals (with the exception of the Amphioxus), as well as in all Arthropoda, this stage of the ciliated larva has been lost, in the course of time, by abbreviated inheritance. There must, however, have existed ancestors of Man in the early Primordial period which possessed the form value of these ciliated larvæ (Planæa, p. 125). A certain proof of this is furnished by the Amphioxus, which is on the one hand related by blood to Man, but on the other has retained down to the present day the stage of the planula.

Fifth Stage: Primæval Stomach Animals (Gastræada).

In the course of the individual development of Amphioxus, as well as in the most different lower animals, there first arises out of the planula the extremely important form of larva which we have named stomach larva, or gastrula (p. 126; Frontispiece, Fig. 5, 6). According to the fundamental law of biogeny this gastrula proves the former existence of an independent form of primæval animal of the same structure, and this we have named primæval stomach animal, or Gastræa (pp. 127, 128). These Gastræada must have existed during the older Primordial period, and they must have also included the ancestors of man. A certain proof of this is furnished by the Amphioxus, which in spite of its blood relationship to Man still passes through the stage of the gastrula with a simple intestine and a double intestinal wall. (Compare Plate X. Fig. B 4.)

Sixth Stage: Gliding Worms (Turbellaria).

The human ancestors of the sixth stage which originated out of the Gastræada of the fifth stage, were low worms, which, of all the forms of worms known to us, were most closely allied to the Gliding Worms, or Turbellaria, or at least upon the whole possessed their form value. Like the Turbellaria of the present day, the whole surface of their body was covered with cilia, and they possessed a simple body of an oval shape, entirely without appendages. These acœlomatous worms did not as yet possess a true body-cavity (cœlom) nor blood. They originated in the early primordial period out of the Gastræada, by the formation of a middle germ-layer, or muscular layer, and also by the further differentiation of the internal parts into various organs; more especially the first formation of a nervous system, the simplest organs of sense, the simplest organs for secretion (kidneys) and generation (sexual organs). The proof that human ancestors existed of a similar formation, is to be looked for in the circumstance that comparative anatomy and ontogeny point to the lower acœlomatous Worms as the common primary form, not merely of all higher Worms, but also of the four higher tribes of animals. Now, of all the animals known to us, the Turbellaria, which possess neither a body-cavity nor blood, are most closely allied to these primæval acœlomatous Primary Worms.283

Seventh Stage: Soft Worms (Scolecida).

Between the Turbellaria of the preceding stage and the Sack Worms of the next stage, we must necessarily assume at least one connecting intermediate stage. For the Tunicata, which of all known animals stand nearest to the eighth stage, and the Turbellaria which most resemble the sixth stage, indeed both belong to the lower division of the unsegmented Worms; but still these two divisions differ so much from one another in their organization, that we must necessarily assume the earlier existence of extinct intermediate forms between the two. These connecting links, of which no fossil remains exist, owing to the soft nature of their bodies, we may comprise as Soft Worms, or Scolecida. They developed out of the Turbellaria of the sixth stage by forming a true body-cavity (a cœlom) and blood in their interior. It is difficult to say which of the still living Cœlomati are nearest akin to these extinct Scolecida; it may be the Acorn-worms (Balanoglossus). The proof that even the direct ancestors of man belonged to these Scolecida, is furnished by the comparative anatomy and the ontogeny of Worms and of the Amphioxus. The form value of this stage must moreover have been represented by several very different intermediate stages, in the wide gap between Turbellaria and Tunicata.

Eighth Stage: Sack Worms (Himatega).

Under the name of Sack worms, or Himatega, we here allude in the eighth place to those Cœlomati, out of which the most ancient skull-less Vertebrata were directly developed. Among the Cœlomati of the present day, the Ascidians are the nearest relatives of these exceedingly remarkable Worms, which connect the widely differing classes of Invertebrate and Vertebrate animals. That the ancestors of man really existed during the primordial period in the form of these Himatega, is distinctly proved by the exceedingly remarkable and important agreement presented by the ontogeny of the Amphioxus and the Ascidia. (Compare Plates XII. and XIII., also pp. 152, 200, etc.) From this fact the earlier existence of Sack Worms may be inferred; they of all known worms were most closely related to our recent Tunicates, especially to the freely swimming young forms or larvæ of the simple Sea-squirts (Ascidia, Phallusia). They originated out of the worms of the seventh stage by the formation of a dorsal nerve-marrow (medulla tube), and by the formation of the spinal rod (chorda dorsalis) which lies below it. It is just the position of this central spinal rod, or axial skeleton, between the dorsal marrow on the dorsal side, and the intestinal canal on the ventral side, which is most characteristic of all Vertebrate animals, including man, but also of the larvæ of the Ascidia. The form value of this stage nearly corresponds with that which the larvæ of the simple Sea-squirts possess at the time when they show the beginning of the dorsal marrow and spinal rod. (Plate XII. Fig. A 5: compare the explanation of these figures in the Appendix.)285

Ninth Stage: Skull-less Animals (Acrania).

The series of human ancestors, which in accordance with their whole organisation we have to consider as Vertebrate animals, begins with the Skull-less animals, or Acrania, of whose nature the still living Lancelet (Amphioxus lanceolatus, Plate XII. B, XIII. B) gives us a faint idea. Since this little animal in its earliest embryonal state entirely agrees with the

Ascidia, and in its further development shows itself to be a true Vertebrate animal, it forms a direct transition from the Vertebrata to the Invertebrata. Even if the human ancestors of the ninth stage in many respects differed from the Amphioxus—the last surviving representative of the Skull-less animals—yet they must have resembled it in its most essential characteristics, in the absence of head, skull, and brain. Skull-less animals of such structure—out of which animals with skulls developed at a later period—lived during the primordial period, and originated out of the Himatega of the eighth stage by the formation of the metamera, or body segments, as also by the further differentiation of all organs, especially the more perfect development of the dorsal nerve-marrow and the spinal rod lying below it. Probably the separation of the two sexes (gonochorism) also began at this stage, whereas all the previously mentioned invertebrate ancestors (apart from the 3—4 first neutral stages) exhibited the condition of hermaphrodites (hermaphroditism). (Compare vol. i. p. 196.) The certain proof of the former existence of these skull-less and brain-less ancestors of man, is furnished by the comparative anatomy and the ontogeny of the Amphioxus and of the Craniota.

Tenth Stage: Single-nostriled Animals (Monorrhina).

Out of the Skull-less ancestors of man there arose in the first place animals with skulls, or Craniota, of the most imperfect nature. The lowest stage of all still living Craniota is occupied by the class of round-mouthed animals, or Cyclostoma, namely, the Hag (Myxinoidea) and Lampreys (Petromyzontia). From the internal organization of these single-nostriled animals, or Monorrhina, we can form an approximate idea of the nature of the human ancestors of the tenth stage. In the former, as also in the latter, skull and brain must have been of the simplest form, and many important organs, as for example, the swimming bladder, the sympathetic nerve, the spleen, the jaw skeleton, and both pairs of legs, may probably as yet not have existed. However, the pouch gills and the round sucking mouth of the Cyclostoma must probably be looked upon as purely adaptive characteristics, which did not exist in the corresponding stage of ancestors. The single-nostriled animals originated during the primordial period out of the skull-less animals by the anterior end of the dorsal marrow developing into the brain, and the anterior end of the dorsal chord into the skull. The certain proof that such single-nostriled and jawless ancestors of man did exist, is found in the "comparative anatomy of the Myxinoidea.

Eleventh Stage: Primæval Fish (Selachii.).

Of all known Vertebrate animals, the ancestors of the Primæval Fish probably showed most resemblance to the still living Sharks (Squalacei). They originated out of the single-nostriled animals by the division of the single nostril into two lateral halves, by the formation of a sympathetic nervous system, a jaw skeleton, a swimming bladder, and two pairs of legs (breast fins or fore-legs, and ventral fins or hind-legs). The internal organisation of this stage may probably, upon the whole, have corresponded to the lowest species of Sharks known to us; the swimming bladder was however more strongly developed; in the case of the latter it exists only as a rudimentary organ. They lived as early as the Silurian period, as is proved by the fossil remains of sharks (teeth and fin spines) from the Silurian strata. A certain proof that the Silurian ancestors of man and of all the other double-nostriled animals were nearest akin to the Selachii, is furnished by the comparative anatomy of the latter; it shows that the relations of organisation in all Amphirrhina can be derived from those of the Selachii.

Twelfth Stage: Mud Fish (Dipneusta).

Our twelfth ancestral stage is formed by Vertebrate animals which probably possessed a remote resemblance to the still living Salamander fish (Ceratodus, Protopterus, Lepidosiren, p. 212). They originated out of the Primæval fish (probably at the beginning of the palæolithic, or primary period) by adaptation to life on land, and by the transformation of the swimming bladder into an air-breathing lung, and of the nasal cavity (which now opened into the cavity of the mouth) into air passages. The series of the ancestors of man which breathed air through lungs began at this stage. Their organisation may probably in many respects have agreed with that of the still living Ceratodus and Protopterus, but at the same time may have been very different. They probably lived at the beginning of the Devonian period. Their existence is proved by comparative anatomy, which shows the Dipneusta to be an intermediate stage between the Selachii and Amphibia.

Thirteenth Stage: Gilled Amphibians (Sozobranchia).

Out of those Mud Fish, which we considered the primary forms of all the Vertebrata which breathe through lungs, there developed the class of Amphibia as the main line (pp. 205, 216). Here began the five-toed formation of the foot (the Pentadactyla), which was thence transmitted to the higher Vertebrata, and finally also to Man. The gilled Amphibians must be looked upon as our most ancient ancestors of the class of Amphibia; besides possessing lungs they retained throughout life regular gills, like the still living Proteus and Axolotl (p. 218). They originated out of the Dipneusta by the transformation of the paddling fins into five-toed legs, and also by the more perfect differentiation of various organs, especially of the vertebral column. In any case they existed about the middle of the palæolithic, or primary period, possibly even before the Coal period; for fossil Amphibia are found in coal. The proof that similar gilled Amphibians were our direct ancestors, is given by the comparative anatomy and the ontogeny of Amphibia and Mammals.289

Fourteenth Stage: Tailed Amphibians (Sozura).

Our amphibious ancestors which retained their gills throughout life, were replaced at a later period by other Amphibia, which, by metamorphosis, lost the gills which they had possessed in early life, but retained the tail, as in the case of the salamanders and newts of the present day. (Compare p. 218.) They originated out of the gilled Amphibians by accustoming themselves in early life to breathe only through gills, and later in life only through lungs. They probably existed even in the second half of the primary, namely, during the Permian period, but possibly even during the Coal period. The proof of their existence lies in the fact that tailed Amphibians form a necessary intermediate link between the preceding and succeeding stages.

Fifteenth Stage: Primæval Amniota (Protamnia).

The name Protamnion we have given to the primary form of the three higher classes of Vertebrate animals, out of which the Proreptilia and the Promammalia developed as two diverging branches (p. 222). It originated out of unknown tailed Amphibia by the complete loss of the gills, by the formation of the amnion, of the cochlea, and of the round window in the auditory organ, and of the organs of tears. It probably originated in the beginning of the mesolithic or secondary period, perhaps even towards the end of the primary, in the Permian period. The certain proof that it once existed lies in the comparative anatomy and the ontogeny of the Amniota; for all Reptiles, Birds, and Mammals, including Man, agree in so many important characteristics that they must, with full assurance, be admitted to be the descendants of a single common primary form, namely, of the Protamnion.

Sixteenth Stage: Primary Mammals (Promammalia).

We now find ourselves more at home with our ancestors. From the sixteenth up to the twenty-second stage they all belong to the large and well known class of Mammals, the confines of which we ourselves have as yet not transgressed. The common, long since extinct and unknown primary forms of all Mammalia, which we have named Promammalia, were at all events, of all still living animals, of the class most closely related to the Beaked animals, or Ornithostoma (Ornithorhynchus, Echidna, p. 233). They differed from the latter, however, by the teeth present in their jaws. The formation of the beak in the Beaked animals of the present day must be looked upon as an adaptive characteristic which developed at a later period. The Promammalia arose out of the Protamnia (probably only at the beginning of the secondary period, namely, in the Trias) by various advances in their internal organisation, as also by the transformation of the epidermal scales into hairs, and by the formation of a mammary gland which furnished milk for the nourishment of the young ones. The certain proof that the Promammalia—inasmuch as they are the common primary forms of all Mammals—also belong to our ancestors, lies in the comparative anatomy and the ontogeny of Mammalia and Man.

Seventeenth Stage: Pouched Animals (Marsupialia).

The three sub-classes of Mammalia—as we have already seen—stand in such a relation to one another that the Marsupials, both as regards their anatomy and their ontogeny and phylogeny, form the direct transition from the Monotrema to Placental animals (p. 247). Consequently, human ancestors must also have existed among Marsupials. They originated out of the Monotrema—which include the primary Mammalia, or Promammalia— by the division of the cloaca into the rectum and the urogenital sinus, by the formation of a nipple on the mammary gland, and by the partial suppression of the clavicles. The oldest Marsupials at all events existed as early as the Jura period (perhaps even in the Trias); during the Chalk period they passed through a series of stages preparing the way for the origin of Placentalia. The certain proof of our derivation from Marsupials—nearly akin to the still living opossum and kangaroo in their essential inner structure—is furnished by the comparative anatomy and the ontogeny of Mammalia.

Eighteenth Stage: Semi-apes (Prosimiæ).

The small group of Semi-apes, as we have already seen, is one of the most important and most interesting orders of Mammalia. It contains the direct primary forms of Genuine Apes, and thus also of Man. Our Semi-ape ancestors probably possessed only a very faint external resemblance to the still living, short-footed Semi-apes (Brachytarsi), especially the Maki, Indri, and Lori (p. 256). They originated (probably at the beginning of the Cenolithic, or Tertiary period) out of Marsupials of Rat-like appearance by the formation of a placenta, the loss of the marsupium and the marsupial bones, and by the higher development of the commissures of the brain. The certain proof that Genuine Apes, and hence also our own race, are the direct descendants of Semi-apes, is to be found in the comparative anatomy and the ontogeny of Placental animals.

Nineteenth Stage: Tailed Apes (Menocerca).

Of the two classes of Genuine Apes which developed out of the Semi-apes, it is only the narrow-nosed, or Catarrhini, which are closely related by blood to Man. Our older ancestors from this group probably resembled the still living Nose-apes and Holy-apes (Semnopithecus), which possess jaws and narrow noses like Man, but have a long tail, and their bodies densely covered with hair (p. 271). The Tailed Apes with narrow noses (Catarrhini Menocerci) originated out of Semi-apes by the transformation of the jaw, and by the claws on their toes becoming changed into nails; this probably took place as early as the older Tertiary period. The certain proof of our derivation from Tailed Catarrhini is to be found in the comparative anatomy and the ontogeny of Apes and of Man.

Twentieth Stage: Man-like Apes (Anthropoides).

Of all still living Apes the large tail-less, narrow-nosed Apes, namely, the Orang and Gibbon in Asia, the Gorilla and Chimpanzee in Africa, are most nearly akin to Man. It is probable that these Man-like Apes, or Anthropoides, originated during the Mid-tertiary period, namely, in the Miocene period. They developed out of the Tailed Catarrhini of the preceding stage—with which they essentially agree—by the loss of the tail, the partial loss of the hairy covering and by the excessive development of that portion of the brain just above the facial portion of the skull. There do not exist direct human ancestors among the Anthropoides of the present day, but they certainly existed among the unknown extinct Human Apes of the Miocene period. The certain proof of their former existence is furnished by the comparative anatomy of Man-like Apes and of Man.

Twenty-first Stage: Ape-like Men (Pithecanthropi).

Although the preceding ancestral stage is already so nearly akin to genuine Men that we scarcely require to assume an intermediate connecting stage, still we can look upon the speechless Primæval Men (Alali) as this intermediate link. These Ape-like men, or Pithecanthropi, very probably existed towards the end of the Tertiary period. They originated out of the Man-like Apes, or Anthropoides, by becoming completely habituated to an upright walk, and by the corresponding stronger differentiation of both pairs of legs. The fore hand of the Anthropoides became the human hand, their hinder hand became a foot for walking. Although these Ape-like Men must not merely by the external formation of their bodies, but also by their internal mental development, have been much more akin to real Men than the Man-like Apes could have been, yet they did not possess the real and chief characteristic of man, namely, the articulate human language of words, the corresponding development of a higher consciousness, and the formation of ideas. The certain proof that such Primæval Men without the power of speech, or Ape-like Men, must have preceded men possessing speech, is the result arrived at by an inquiring mind from comparative philology (from the "comparative anatomy" of language), and especially from the history of the development of language in every child ("glottal ontogenesis") as well as in every nation ("glottal phylogenesis").

Twenty-second Stage: Men (Homines).

Genuine Men developed out of the Ape-like Men of the preceding stage by the gradual development of the animal language of sounds into a connected or articulate language, of words. The development of this function, of course, went hand in hand with the development of its organs, namely, the higher differentiation of the larynx and the brain. The transition from speechless Ape-like Men to Genuine or Talking Men probably took place at the beginning of the Quaternary period, namely, in the Diluvial period, but possibly even at an earlier date, in the more recent Tertiary. As, according to the unanimous opinion of most eminent philologists, all human languages are not derived from a common primæval language, we must assume a polyphyletic origin of language, and in accordance with this a polyphyletic transition from speechless Ape-like Men to Genuine Men.

MIGRATION AND DISTRIBUTION OF MANKIND

The rich treasure of knowledge we possess in the comparative anatomy and the history of the development of Vertebrate animals, enables us even now to establish the most important outlines of the human pedigree in the way we have done in the last chapter. One must, however, not expect to be able to survey satisfactorily in every detail the history or phylogeny of the human species which will henceforth form the basis of Anthropology, and of all other sciences. The complete development of this most important science—of which we can only lay the first foundation—must remain reserved for the more accurate and extensive investigations of a future time. This applies also to those more special questions of human phylogeny at which it is desirable before concluding to take a cursory glance, namely, the question of the time and place of the origin of the human race, as also of the different species and races into which it has differentiated.

In the first place, the period of the earth's history, within which the slow and gradual transmutation of the most man-like apes into the most ape-like men took place, can of course not be determined by years, nor even by centuries. This much can, however, with full assurance be maintained, for reasons given in the last chapter, that Man is derived from Placental animals. Now, as fossil remains of these Placentalia are found only in the tertiary rocks, the human race can at the earliest have developed only within the Tertiary period out of perfected man-like apes. What seems most probable is that this most important process in the history of terrestrial creation occurred towards the end of the Tertiary period, that is in the Pliocene, perhaps even in the Miocene period, but possibly also not until the beginning of the Diluvial period. At all events Man, as such, lived in central Europe as early as the Diluvial period, contemporaneously with many large, long since extinct mammals, especially with the diluvial elephant, or mammoth (Elephas primigenius), the woolly-haired rhinoceros (Rhinoceros tichorrhinus), the giant deer (Cervus euryceros), the cave bear (Ursus spelæus), the cave hyæna (Hyæna spelæa), the cave lion (Felis spelæus), etc. The results brought to light by recent geology and archæology as to these fossil men and their animal contemporaries of the diluvial period, are of the greatest interest. But as a closer examination of them would occupy too much of my limited space, I must confine myself here to setting forth their great general importance, and refer for particulars to the numerous writings which have recently been published on the Primæval History of Man, more especially to the excellent works of Charles Lyell,(30) Carl Vogt,(27) Friedrich Rolle,(28) John Lubbock,(44) L. Büchner,(43) etc.

The numerous and interesting discoveries presented to us by these extensive investigations of late years on the primæval history of the human race, place the important fact (long since probable for many other reasons) beyond a doubt, that the human race, as such, has existed for more than twenty thousand years. But it is also probable that more than a hundred thousand years, perhaps many hundred thousands of years, have elapsed since its first appearance; and, in contrast to this, it must seem very absurd that our calendars still represent the "Creation of the World, according to Calvisius," to have taken place 5821 years ago.

Now, whether we reckon the period during which the human race, as such, has existed and diffused itself over the earth, as twenty thousand, a hundred thousand, or many hundred thousands of years, the lapse of time is in any case immensely small in comparison with the inconceivable length of time which was requisite for the gradual development of the long chain of human ancestors. This is evident even from the small thickness of all Diluvial deposits in comparison with the Tertiary, and of these again in comparison with the preceding deposits. (Compare p. 22.) But the infinitely long series of slowly and gradually developing animal forms from the simplest Moneron to the Amphioxus, from this to the Primæval Fish, from the Primæval Fish to the first Mammal, and again, from the latter to Man, also require for their historical development a succession of periods probably comprising many thousands of millions of years. (Compare vol. i. p. 129.)

Those processes of development which led to the origin of the most Ape-like Men out of the most Man-like Apes must be looked for in the two adaptational changes which, above all others, are distinctive of Man, namely, upright walk and articulate speech. These two physiological functions necessarily originated together with two corresponding morphological transmutations, with which they stand in the closest correlation, namely, the differentiation of the two pairs of limbs and the differentiation of the larynx. The important perfecting of these organs and their functions must have necessarily and powerfully reacted upon the differentiation of the brain and the mental activities dependent upon it, and thus have paved the way for the endless career in which Man has since progressively developed, and in which he has far outstripped his animal ancestors. (Gen. Morph. ii. p. 430.)

The first and earliest of these three great processes in the development of the human organism probably was the higher differentiation and the perfecting of the extremities which was effected by the habit of an upright walk. By the fore feet more and more exclusively adopting and retaining the function of grasping and handling, and the hinder feet more and more exclusively the function of standing and walking, there was developed that contrast between the hand and foot which is indeed not exclusively characteristic of man, but which is much more strongly developed in him than in the apes most like men. This differentiation of the fore and hinder extremities was, however, not merely most advantageous for their own development and perfecting, but it was followed at the same time by a whole series of very important changes in other parts of the body. The whole vertebral column, and more especially the girdle of the pelvis and shoulders, as also the muscles belonging to them, thereby experienced those changes which distinguish the human body from that of the most man-like apes. These transmutations were probably accomplished long before the origin of articulate speech; and the human race thus existed for long, with an upright walk and the characteristic human form of body connected with it, before the actual development of human language, which would have completed the second and the more important part of human development. We may therefore distinguish a special (21st) stage in the series of our human ancestors, namely, Speechless Man (Alalus), or Ape-man (Pithecanthropus), whose body was indeed formed exactly like that of Man in all essential characteristics, but who did not as yet possess articulate speech.

The origin of articulate language, and the higher differentiation and perfecting of the larynx connected with it, must be looked upon as only a later, and the most important stage in the process of the development of Man. It was, doubtless, this process which above all others helped to create the deep chasm between man and animal, and which also first caused the most important progress in the mental activity and the perfecting of the brain connected with it. There indeed exists in very many animals a language for communicating sensations, desires, and thoughts, partly a language of gestures, partly a language of feeling or touch, partly a language of cries or sounds, but a real language of words or ideas, a so-called "articulate" language, which by abstraction changes sounds into words, and words into sentences, belongs, as far as we know, exclusively to Man.

The origin of human language must, more than anything else, have had an ennobling and transforming influence upon the mental life of Man, and consequently upon his brain. The higher differentiation and perfecting of the brain and mental life as its highest function developed in direct correlation with its expression by means of speech. Hence, the highest authorities in comparative philology justly see in the development of human speech the most important process which distinguishes Man from his animal ancestors. This has been especially set forth by August Schleicher, in his treatise "On the Importance of Speech for the Natural History of Man."(34) In this relation we see one of the closest connections between comparative zoology and comparative philology; and here the theory of development assigns to the latter the task of following the origin of language step by step. This task, as interesting as it is important, has of late years been successfully undertaken by many inquirers, but more especially by Wilhelm Bleek, who has been occupied for seventeen years in South Africa with the study of the languages of the lowest races of men, and hence has been enabled to solve the question. August Schleicher more especially discusses, in accordance with the theory of selection, how the various forms of speech, like all other organic forms and functions, have developed by the process of natural selection, and have divided into many species and dialects.

I have no space here to follow the process of the formation of language, and must refer in regard to this to the above-mentioned important work of Wilhelm Bleek, "On the Origin of Language."(35) But we have still to mention one of the most important results of comparative philology, which is of the highest importance to the genealogy of the human species, that is, that human language was probably of a multiple, or polyphyletic origin. Human speech, as such, did not develop probably until the genus of Speech-less or Primæval Man, or Ape Man, had separated into several kinds or species. In each of these human species, and perhaps even in the different sub-species and varieties of this species, language developed freely and independently of the others. At least Schleicher, one of the first authorities on the subject, maintains that "even the beginnings of language—in sounds as well as in regard to ideas and views which were reflected in sounds, and further, in regard to their capability of development—must have been different. For it is positively impossible to trace all languages to one and the same primæval language. An impartial investigation rather shows that there are as many primæval languages as there are races."(34) In like manner, Friederich Müller(41) and other eminent linguists assume a free and independent origin of the families of languages and their primæval stocks. It is well known, however, that the boundaries of these tribes of languages and their ramifications are by no means always the boundaries of the different human species, or the so-called "races," distinguished by us on account of their bodily character 303istics. This, as well as the complicated relations of the mixture of races, and the various forms of hybrids, is the great difficulty lying in the way of tracing the human pedigree in its individual branches, species, races, varieties, etc.

In spite of these great and serious difficulties, we cannot here refrain from taking one more cursory glance at the ramification of the human pedigree, and at the same time considering, from the point of view of the theory of descent, the much discussed question of the monophyletic or polyphyletic origin of the human race, and its species or races. As is well known, two great parties have for a long time been at war with each other upon this question; the monophylists (or monogenists) maintain the unity of origin and the blood relationship of all races of men. The polyphylists (or polygenists), on the other hand, are of opinion that the different races of men are of independent origin. According to our previous genealogical investigations we cannot doubt that, at least in a wide sense, the monophyletic opinion is the right one. For even supposing that the transmutation of Man-like Apes into Men had taken place several times, yet those Apes themselves would again be allied by the one pedigree common to the whole order of Apes. The question therefore would always be merely about a nearer or remoter degree of blood relationship. In a narrower sense, on the other hand, the polyphylist's opinion would probably be right, inasmuch as the different primæval languages have developed quite independently of one another. Hence, if the origin of an articulate language is considered as the real and principal act of humanification, and the species of the human race are distinguished according to the roots of their language, it might be said that the different races of men had originated, independently of one another, by different branches of primæval, speechless men directly springing from apes, and forming their own primæval language. Still they would of course be connected further up or lower down at their root, and thus all would finally be derived from a common primæval stock.

While we hold the latter of these convictions, and while we for many reasons believe that the different species of speechless primæval men were all derived from a common ape-like human form, we do not of course mean to say that all men are descended from one pair. This latter supposition, which our modern Indo-Germanic culture has taken from the Semitic myth of the Mosaic history of creation, is by no means tenable. The whole of the celebrated dispute, as to whether the human race is descended from a single pair or not, rests upon a completely false way of putting the question. It is just as senseless as the dispute as to whether all sporting dogs or all race-horses are descended from a single pair. We might with equal justice ask whether all Germans or all Englishmen are "descended from a single pair," etc. A "first human pair," or "a first man," has in fact never existed, any more than there ever existed a first pair or a first individual of

Englishmen, Germans, race-horses, or sporting dogs. The origin of a new species, of course, always results from an existing species, by a long chain of many different individuals sharing the slow process of transformation. Supposing that we had all the different pairs of Human Apes and Ape-like Men before us—which belong to the true ancestors of the human race—it would even then be quite impossible (without doing so most arbitrarily) to call any one of these pairs of ape-like men "the first pair." As little can we derive each of the twelve races or species of men, which we shall consider directly, from a "first pair."

The difficulties met with in classifying the different races or species of men are quite the same as those which we discover in classifying animal and vegetable species. In both cases forms apparently quite different are connected with one another by a chain of intermediate forms of transition. In both cases the dispute as to what is a kind or a species, what a race or a variety, can never be determined. Since Blumenbach's time, as is well known, it has been thought that mankind may be divided into five races or varieties, namely: (1) the Ethiopian, or black race (African negro); (2) the Malayan, or brown race (Malays, Polynesians, and Australians); (3) the Mongolian, or yellow race (the principal inhabitants of Asia and the Esquimaux of North America); (4) the Americans, or red race (the aborigines of America); and (5) the Caucasian, or white race (Europeans, north Africans, and south-western Asiatics). All of these five races of men, according to the Jewish legend of creation, are said to have been descended from "a single pair"—Adam and Eve,—and in accordance with this are said to be varieties of one kind or species. If, however, we compare them without prejudice, there can be no doubt that the differences of these five races are as great and even greater than the "specific differences" by which zoologists and botanists distinguish recognised "good" animal and vegetable species ("bonæ species"). The excellent palæontologist Quenstedt is right in maintaining that, "if Negroes and Caucasians were snails, zoologists would universally agree that they represented two very excellent species, which could never have originated from one pair by gradual divergence."

The characteristics by which the races of men are gradually distinguished are partly taken from the formation of the hair, partly from the colour of the skin, and partly from the formation of the skull. In regard to the last character, two extremes are distinguished, namely, long heads and short heads. In long-headed men (Dolichocephali), whose strongest development is found in Negroes and Australians, the skull is extended, narrow, and compressed on the right and left. In short-headed men (Brachycephali), on the other hand, the skull is compressed in an exactly opposite manner, from the front to the back, is short and broad, which is especially striking in the case of the Mongolians. Medium-headed men (Mesocephali), standing between the two extremes, predominate especially among Americans. In every one of these three groups we find men with slanting teeth (Prognathi), whose jaws, like those of the animal snout, strongly project, and whose front teeth therefore slope in front, and men with straight teeth (Orthognathi), whose jaws project but little, and whose front teeth stand perpendicularly. During the last ten years a great deal of time and trouble have been devoted to the careful examination and measurement of the forms of skulls, which have, however, not been rewarded by corresponding results. For within a single species, as for example within the Mediterranean species, the form of the skull may vary so much that both extremes are met with in the same species. Much better starting-points for the classification of of the human species are furnished by the nature of the hair and speech, because they are much more strictly hereditary than the form of the skull.

Comparative philology seems especially to be becoming an authority in this matter. In the latest great work on the races of men, which Friederich Müller has published in his excellent "Ethnography,"(42) he justly places language in the fore-ground. Next to it the nature of the hair of the head is of great importance; for although it is in itself of course only a subordinate morphological character, yet it seems to be strictly transmitted within the race. Of the twelve species of men distinguished on the following table (p. 308), the four lower species are characterised by the woolly nature of the hair of their heads; every hair is flattened like a tape, and thus its section is oval. These four species of woolly-haired men (Ulotrichi) we may reduce into two groups—tuft-haired and fleecy-haired. The hair on the head of tuft-haired men (Lophocomi), Papuans and Hottentots, grows in unequally divided small tufts. The woolly hair of fleecy-haired men (Eriocomi), on the other hand, in Caffres and Negroes, grows equally all over the skin of the head. All Ulotrichi, or woolly-haired men, have slanting teeth and long heads, and the colour of their skin, hair, and eyes is always very dark. All are inhabitants of the Southern Hemisphere; it is only in Africa that they come north of the equator. They are on the whole at a much lower stage of development, and more like apes, than most of the Lissotrichi, or straight-haired men. The Ulotrichi are incapable of a true inner culture and of a higher mental development, even under the favourable conditions of adaptation now offered to them in the United States of North America. No woolly-haired nation has ever had an important "history."

In the eight higher races of men, which we comprise as straight-haired (Lissotrichi), the hair of the head is never actually woolly, although it is very much frizzled in some individuals. Every separate hair is cylindrical (not like a tape), and hence its section is circular (not oval).

The eight races of Lissotrichi may likewise be divided into two groups— stiff-haired and curly-haired. Stiff-haired men (Euthycomi), the hair of whose heads is quite smooth and straight, and not frizzled, include

Australians, Malays, Mongolians, Arctic tribes, and Americans. Curly-haired men, on the other hand, the hair of whose heads is more or less curly, and in whom the beard is more developed than in all other species, include the Dravidas, Nubians, and Mediterranean races. (Compare Plate XV.)

Now, before we venture upon the attempt hypothetically to explain the phyletic divergence of mankind, and the genealogical connection of its different species, we will premise a short description of the twelve named species and of their distribution. In order clearly to survey their geographical distribution, we must go back some three or four centuries, to the time when the Indian Islands and America were first discovered, and when the present great mingling of species, and more especially the influx of the Indo-Germanic race, had as yet not made great progress. We begin with the lowest stages, with the woolly-haired men (Ulotrichi), all of whom are prognathic Dolichocephali.

The Papuan (Homo Papua), of all the still living human species, is perhaps most closely related to the original primary form of woolly-haired men. This species now inhabits only the large island of New Guinea and the Archipelago of Melanesia lying to the east of it (Solomon's Islands, New Caledonia, the New Hebrides, etc.). But scattered remnants of it are also still found in the interior of the peninsula of Malacca, and likewise in many other islands of the large Pacific Archipelago; mostly in the inaccessible mountainous parts of the interior, and especially in the Philippine Islands. The but lately extinct Tasmanians, or the natives of Van Diemen's Land, belonged to this group. From these and other circumstances it is clear that the Papuans in former times possessed a much larger area of distribution in south-eastern Asia. They were driven out by the Malays and forced eastwards. The skin of all Papuans is of a black colour, sometimes more inclining to brown, sometimes more to blue. Their woolly hair grows in tufts, is spirally twisted in screws, and often more than a foot in length, so that it forms a strong woolly wig, which stands far out from the head. Their face, below the narrow depressed forehead, has a large turned-up nose and thick protruding lips. The peculiar form of their hair and speech so essentially distinguishes the Papuans from their straight-haired neighbours, from the Malays as well as from the Australians, that they must be regarded as an entirely distinct species.

Closely related to the Papuans by the tufted growth of hair, but geographically widely separated from them, are the Hottentots (Homo Hottentottus). They inhabit exclusively the southernmost part of Africa, the Cape and the adjacent parts, and have immigrated there from the north-east. The Hottentots, like their original kinsmen the Papuans, occupied in former times a much larger area (probably the whole of Eastern Africa), and are now approaching their extinction. Besides the genuine Hottentots—of whom there now exist only the two tribes of the Coraca (in the eastern Cape districts) and the Namaca (in the western portion of the Cape)—this species also includes the Bushmen (in the mountainous interior of the Cape). The woolly hair of all Hottentots grows in tufts, like brushes, as in the case of Papuans. Both species also agree in the posterior part of the body, in the female sex being specially inclined to form a great accumulation of fat (Steatopygia). But the skin of Hottentots is much lighter, of a yellowish brown colour. Their very flat face is remarkable for its small forehead and nose, and large nostrils. The mouth is very broad with big lips, the chin small and pointed. Their speech is characterised by several quite peculiar guttural sounds.

The next neighbours and kinsmen of Hottentots are Kaffres (Homo Cafer). This woolly-haired human species is, however, distinguished, like the following one (the genuine Negro), from Hottentots and Papuans by the woolly hair not being divided into tufts, but covering the head as a thick fleece. The colour of their skin varies through all shades, from the yellowish black of the Hottentot to the brown black or pure black of the genuine Negro. While in former times the race of Kaffres was assigned to a very small area of distribution, and was generally looked upon only as a variety of the genuine Negro, this species is now considered to include almost the whole of the inhabitants of equatorial Africa, from the 20th degree south latitude to the 4th degree north; consequently, all South Africans, with the exception of the Hottentots. They include especially the inhabitants of the Zulu, Zambesi, and Mozambique districts on the east coast, the large human families of the Beschuans or Setschuans in the interior, and the Herrero and Congo tribes of the west coast. They too, like the Hottentots, have immigrated from the north-east. Kaffres, who were usually classed with Negroes, differ very essentially from them by the formation of their skull and by their speech. Their face is long and narrow, their forehead high, and their nose prominent and frequently curved, their lips not so protruding, and their chin pointed. The many languages of the different tribes of Kaffres can all be derived from an extinct primæval language, namely, from the Bantu language.

The genuine Negro (Homo Niger)—when Kaffres, Hottentots, and Nubians are separated from him—at present forms a much less comprehensive human species than was formerly supposed. They now only include the Tibus, in the eastern parts of the Sahara; the Sudan people, or Sudians, who inhabit the south of that large desert; also the inhabitants of the Western Coast of Africa, from the mouth of the Senegal in the north, to beyond the estuary of the Niger in the south (Senegambians and Nigritians). Genuine Negroes are accordingly confined between the equator and the Tropic of Capricorn, and only a small portion of the Tibu tribe in the east have gone beyond this boundary. The Negro species has spread within this zone, coming from the east. The colour of the skin of genuine negroes is always more or less of a pure black. Their skin is velvety to the touch, and characterised by a peculiar offensive exhalation. Although Negroes agree with Kaffres in the formation of the woolly hair of the head, yet they differ essentially in the formation of their face. Their forehead is flatter and lower, their nose broad and thick, not prominent, their lips large and protruding, and their chin very short. Genuine Negroes are moreover distinguished by very thin calves and very long arms. This species of men must have branched into many separate tribes at a very early period, for their numerous and entirely distinct languages can in no way be traced to one primæval language.

To the four woolly-haired species of men just discussed, straight-haired men (Homines Lissotrichi) stand in strong contrast, as another main branch of the genus. Five of the eight species of the latter, as we have seen, can be comprised as stiff-haired (Euthycomi) and three as curly-haired (Euplocomi). We shall in the first place consider the former, which includes the primæval inhabitants of the greater part of Asia and the whole of America.

The lowest stage of all straight-haired men, and on the whole perhaps of all the still living human species, is occupied by the Australian, or Austral-negro (Homo Australis). This species seems to be exclusively confined to the large island of Australia; it resembles the genuine African Negro by its black or brownish black hair, and the offensive smell of the skin, by its very slanting teeth and long-headed form of skull, the receding forehead, broad nose, protruding lips, and also by the entire absence of calves. On the other hand Australians differ from genuine Negroes as well as from their nearest neighbours the Papuans, by the much weaker and more delicate structure of their bones, and more especially by the formation of the hair of their heads, which is not woolly and frizzled, but either quite lank or only slightly curled. The very low stage of bodily and mental development of the Australian is perhaps not altogether original, but has arisen by degeneration, that is, by adaptation to the very unfavourable conditions of existence in Australia. They probably immigrated to their present home from the north or north-west, as a very early offshoot of the Euthycomi. They are probably more closely related to the Dravidas, and hence to the Euplocomi, than the other Euthycomi. The very peculiar language of the Australians is broken up into numerous small branches, which are grouped into a northern and a southern class.

The Malay (Homo Malayus), the brown race of ethnographers, although not a large species, is important in regard to its genealogy. An extinct south Asiatic human species, very closely related to the Malays of the present day, must probably be looked upon as the common primary form of this and the following higher human species. We will call this hypothetical primary species, Primæval Malays, or Promalays. The Malays of the present day are divided into two widely dispersed races, the Sundanesians, who inhabit Malacca, the Sunda Islands (Sumatra, Java, Borneo, etc.) and the Philippine Islands, and the Polynesians, who are dispersed over the greater portion of the Pacific Archipelago. The northern boundary of their wide tract of distribution is formed on the east by the Sandwich Islands (Hawai), and on the west by the Marian Islands (Ladrones); the southern boundary on the east is formed by the Mangareva Archipelago, and on the west by New Zealand. The inhabitants of Madagascar are an especial branch of Sundanesians who have been driven to the far west. This wide pelagic distribution of the Malays is explained by their partiality for nautical life. Their primæval home is the south-eastern portion of the Asiatic continent, from whence they spread to the east and south, and drove the Papuans before them. The Malays, in the formation of body, are nearest akin to the Mongols, but are also nearly allied to the curly-haired Mediterranese. They are generally short-headed, more rarely medium-headed, and very rarely long-headed. Their hair is black and stiff, but frequently somewhat curled. The colour of their skin is brown, sometimes yellowish, or of a cinnamon colour, sometimes reddish or copper brown, more rarely dark brown. In regard to the formation of face, Malays in a great measure form an intermediate stage between the Mongols and the Mediterranese; they can frequently not be distinguished from the latter. Their face is generally broad, with prominent nose and thick lips, the opening for their eyes not so narrowly cut and slanting as in Mongols. The near relationship between all Malays and Polynesians is proved by their language, which indeed broke up at an early period into many small branches, but still can always be traced to a common and quite peculiar primæval language.

The Mongol (Homo Mongolus) is, next to the Mediterranese, the richest in individuals. Among them are all the inhabitants of the Asiatic Continent, excepting the Hyperboreans in the north, the few Malays in the south-east (Malacca), the Dravidas in Western India, and the Mediterranese in the south-west. In Europe this species of men is represented by the Fins and Lapps in the north, by the Osmanlis in Turkey, and the Magyars in Hungary. The colour of the Mongol is always distinguished by a yellow tone, sometimes a light pea green, or even white, sometimes a darker brownish yellow. Their hair is always stiff and black. The form of their skull is, in the great majority of cases, decidedly short (especially in Kalmucks, Baschkirs, etc.) but frequently of medium length (Tartars, Chinese, etc.) But among them we never meet with genuine long-headed men. The narrow openings of their eyes, which are generally slanting, their prominent cheek bones, broad noses, and thick lips are very striking, as well as the round form of their faces. The language of the Mongols is probably traceable to a common primæval language; but the monosyllabic languages of the Indo-Chinese races, and the polysyllabic languages of the other

Mongol races, stand in contrast as two main branches which separated at an early time. The monosyllabic tribes of the Indo-Chinese include the Tibetans, Birmans, Siamese, and Chinese. The other polysyllabic Mongols are divided into three races, namely: (1) the Coreo-Japanese (Coreans and Japanese); (2) the Altaians (Tartars, Kirgises, Kalmucks, Buriats, Tungusians); and (3) the Uralians (Samoiedes, Fins). The Magyars of Hungary are descended from the Fins.

The Polar men (Homo Arcticus) must be looked upon as a branch of the Mongolian human species. We comprise under this name the inhabitants of the Arctic Polar lands of both hemispheres, the Esquimaux (and Greenlanders) in North America, and the Hyperboreans in north-eastern Asia (Jukagirs, Tschuksches, Kuriaks, and Kamtschads). By adaptation to the Polar climate, this human race has become so peculiarly transformed that it may be considered as a distinct species. Their stature is low and of a square build; the formation of their skull of medium size or even long; their eyes narrow and slanting like the Mongols; their cheek-bones prominent, and their mouth wide. Their hair is stiff and black; the colour of their skin is of a light or dark brown tinge, sometimes more inclined to white or to yellow, like that of the Mongols, sometimes more to red, like that of the Americans. The languages of Polar men are as yet little known, but they differ both from the Mongolian and from the American. Polar men must probably be regarded as a remnant and a peculiarly adapted branch of that tribe of Mongols which emigrated from north-eastern Asia to North America, and populated that part of the earth.

At the time of the discovery of America, that part of the earth was peopled (setting aside the Esquimaux) only by a single human species, namely, by the Redskins, or Americans (Homo Americanus). Of all other human species they are most closely related to the two preceding. The form of their skull is generally a medium one, rarely short or long-headed. Their forehead broad and very low; their nose large, prominent, and frequently aquiline; their cheek-bones prominent; their lips rather thin than thick. The colour of their skin is characterised by a red fundamental tint, which is, however, sometimes pure copper-red, or light red, sometimes a deeper reddish brown, yellow brown or olive brown. The numerous languages of the various American races and tribes are extremely different, yet they agree in their original foundation. Probably America was first peopled from north-eastern Asia by the same tribe of Mongols from whom the Polar men (Hyperboreans and Esquimaux) have also branched. This tribe first spread in North America, and from thence migrated over the isthmus of Central America down to South America, at the extreme south of which the species degenerated very much by adaptation to the very unfavourable conditions of existence. But it is also possible that Mongols and Polynesians immigrated from the west and mixed with the former tribe. In any case the aborigines of America came over from the Old World, and did not, as some suppose, in any way originate out of American apes. Catarrhini, or Narrow-nosed Apes, never at any period existed in America.

The three human species still to be considered—the Dravidas, Nubians, and Mediterranese—agree in several characteristics which seem to establish a close relationship between them, and distinguish them from the preceding species. The chief of these characteristics is the strong development of the beard, which in all other species is either entirely wanting or but very scanty. The hair of their heads is generally not so lank and smooth as in the five preceding species, but in most cases more or less curly. Other characteristics also seem to favour our classing them in one main group of curly-haired men (Euplocomi).

The Dravida man (Homo Dravida) seems to stand very near the common primary form of the Euplocomi, and perhaps of Lissotrichi. At present this primæval species is only represented by the Deccan tribes in the southern part of Hindostan, and by the neighbouring inhabitants of the mountains on the north-east of Ceylon. But in earlier times this race seems to have occupied the whole of Hindostan, and to have spread even further. It shows, on the one hand, traits of relationship to the Australians and Malays; on the other, to the Mongols and Mediterranese. Their skin is either of a light or dark brown colour; in some tribes, of a yellowish brown, in others, almost black brown. The hair of their heads, as in Mediterranese, is more or less curled, neither quite smooth, like that of the Euthycomi, nor actually woolly, like that of the Ulotrichi. The strong development of the beard is also like that of the Mediterranese. The oval form of face seems partly to be akin to that of the Malays, partly to that of the Mediterranese. Their forehead is generally high, their nose prominent and narrow, their lips slightly protruding. Their language is now very much mixed with Indo-Germanic elements, but seems to have been originally derived from a very peculiar primæval language.

The Nubian (Homo Nuba) has caused ethnographers no fewer difficulties than the Dravida species. By this name we understand not merely the real Nubians (Schangallas, or Dongolese), but also their near kinsmen, the Fulas, or Fellatas. The real Nubians inhabit the countries of the Upper Nile (Dongola, Schangalla, Barabra, Cordofan); the Fulas, or Fellatas, on the other hand, have thence migrated far westward, and now inhabit a broad tract in the south of the western Sahara, hemmed in between the Soudanians in the north and the Nigritos in the south. The Nubian and Fula races are generally either classed with negroes or with the Hamitic races (thus with Mediterranese), but are so essentially different from both that they must be regarded as a distinct species. In former times they very probably occupied a large part of north-eastern Africa. The skin of the Nubian and Fula races is of a yellowish or reddish brown colour, more rarely dark brown or approaching to black. Their hair is not woolly but curled, frequently even quite smooth; its colour is dark brown or black. Their beard is much more strongly developed than in negroes. The oval formation of their faces approaches more to the Mediterranean than to the Negro type. Their forehead is high and broad, their nose prominent and not flat, their lips not so protruding as in the negro. The language of the Nubian races seems to possess no relationship to those of genuine negroes.

The Caucasian, or Mediterranean man (Homo Mediterraneus), has from time immemorial been placed at the head of all races of men, as the most highly developed and perfect. It is generally called the Caucasian race, but as among all the varieties of the species, the Caucasian branch is the least important, we prefer the much more suitable appellation proposed by Friedrich Müller, namely, that of Mediterranean, or Midland men. For the most important varieties of this species, which are moreover the most eminent actors in what is called "Universal History," first rose to a flourishing condition on the shores of the Mediterranean. The former area of the distribution of this species is expressed by the name of "Indo-Atlantic" species, whereas at present it is spread over the whole earth, and is overcoming most of the other species in the struggle for existence. In bodily as well as in mental qualities, no other human species can equal the Mediterranean. This species alone (with the exception of the Mongolian) has had an actual history; it alone has attained to that degree of civilization which seems to raise man above the rest of nature.

The characteristics which distinguish the Mediterranean from the other species of the race are well known. The chief of the external features is the light colour of the skin, which however exhibits all shades, from pure white or reddish white, through yellow or yellowish brown to dark brown or even black brown. The growth of the hair is generally strong, the hair of the head more or less curly, the hair of the beard stronger than in any of the other species. The form of the skull shows a great development in breadth; medium heads predominate upon the whole, but long and short heads are also widely distributed. It is only in this one species of men that the body as a whole attains that symmetry in all parts, and that equal development, which we call the type of perfect human beauty. The languages of all the races of this species can by no means be traced to a single common primæval language; we must at least assume four radically different primæval languages. In accordance with this we must also assume within this one species four different races, which are only connected at their root. Two of these races, the Basques and Caucasians, now exist only as small remnants. The Basques, which in earlier times peopled the whole of Spain and the south of France, now inhabit but a narrow tract of land on the northern coast of Spain, on the Bay of Biscay. The remnant of the Caucasian race (the Daghestans, Tschercassians, Mingrelians, and

Georgians) are now confined to the districts of Mount Caucasus. The language of the Caucasians as well as that of the Basques is entirely peculiar, and can be traced neither to the Semitic nor to the Indo-Germanic primæval languages.

Even the languages of the two principal races of the Mediterranean species—the Semitic and Indo-Germanic—cannot be traced to a common origin, and consequently these two races must have separated at a very early period. Semites and Indo-Germani are descended from different ape-like men. The Semitic race likewise separated at a very early period into two diverging branches, namely, into the Egyptian and Arabic branches. The Egyptian, or African branch, the Dyssemites—which sometimes under the name of Hamites are entirely separated from the Semites—embraces the large group of Berbers, who occupy the whole of north Africa, and in earlier times also peopled the Canary Islands, and, finally, also the group of the Ethiopians, the Bedsha, Galla, Danakil, Somali, and other tribes which occupy all the north-eastern shores of Africa as far as the equator. The Arabic, or Asiatic branch, that is, the Eusemites, also called Semites in a narrow sense, embrace the inhabitants of the large Arabian peninsula, the primæval family of genuine Arabians ("primæval type of the Semites"), and also the most highly developed Semitic groups, the Jews, or Hebrews, and the Aramæans—the Syrians and Chaldæans. A colony of the southern Arabs (the Himjarites), which crossed the Straits of Bab-el-Mandeb, has peopled Abyssinia.

Lastly, the Indo-Germanic race, which has far surpassed all the other races of men in mental development, separated at a very early period, like the Semitic, into two diverging branches, the Ario-Romaic and the Slavo-Germanic branches. Out of the former arose on the one hand the Arians (Indians and Iranians), on the other the Græco-Roman (Greeks and Albanians, Italians and Kelts). Out of the Slavo-Germanic branch were developed on the one hand the Slavonians (Russian, Bulgarian, Tchec, and Baltic tribes), on the other the Germani (Scandinavians and Germans, Netherlanders and Anglo-Saxons). August Schleicher has explained, in a very clear genealogical form, how the further ramifications of the Indo-Germanic race may be accurately traced in detail on the basis of comparative philology.(6) (Compare p. 331.)

The total number of human individuals at present amounts to between 1,300 and 1,400 millions. In our Tabular Survey (p. 333) 1,350 millions has been assumed as the mean number. According to an approximate estimate, as far as such a thing is possible, 1,200 millions of these are straight-haired men, only about 150 millions woolly-haired. The most highly developed species, Mongols and Mediterranese, far surpass all the other human species in numbers of individuals, for each of them alone comprises about 550 millions. (Compare Friederich Müller's Ethnography, p. 30.) Of course the relative number of the twelve species fluctuates every year, and that too according to the law developed by Darwin, that in the struggle for life the more highly developed, the more favoured and larger groups of forms, possess the positive inclination and the certain tendency to spread more and more at the expense of the lower, more backward, and smaller groups. Thus the Mediterranean species, and within it the Indo-Germanic, have by means of the higher development of their brain surpassed all the other races and species in the struggle for life, and have already spread the net of their dominion over the whole globe. It is only the Mongolian species which can at all successfully, at least in certain respects, compete with the Mediterranean. Within the tropical regions, Negroes, Kaffres, and Nubians, as also the Malays and Dravidas, are in some measure protected against the encroachments of the Indo-Germanic tribes by their being better adapted for a hot climate; the case of the arctic tribes of the polar regions is similar. But the other races, which as it is are very much diminished in number, will sooner or later completely succumb in the struggle for existence to the superiority of the Mediterranean races. The American and Australian tribes are even now fast approaching their complete extinction, and the same may be said of the Papuans and Hottentots.

In now turning to the equally interesting and difficult question of the relative connection, migration, and primæval home of the twelve species of men, I must premise the remark that, in the present state of our anthropological knowledge, any answer to this question must be regarded only as a provisional hypothesis. This is much the same as with any genealogical hypothesis which we may form of the origin of kindred animal and vegetable species, on the basis of the "Natural System." But the necessary uncertainty of these special hypotheses of descent, in no way shakes the absolute certainty of the general theory of descent. Man, we may feel certain, is descended from Catarrhini, or narrow-nosed apes, whether we agree with the polyphylites, and suppose each human species, in its primæval home, to have originated out of a special kind of ape; or whether, agreeing with the monophylites, we suppose that all the human species arose only by differentiation from a single species of primæval man (Homo primigenius).

For many and weighty reasons we hold the monophyletic hypothesis to be the more correct, and we therefore assume a single primæval home for mankind, where he developed out of a long since extinct anthropoid species of ape. Of the five now existing continents, neither Australia, nor America, nor Europe can have been this primæval home, or the so-called "Paradise," the "cradle of the human race." Most circumstances indicate southern Asia as the locality in question. Besides southern Asia, the only other of the now existing continents which might be viewed in this light is Africa. But there are a number of circumstances (especially chorological facts) which suggest that the primæval home of man was a continent now sunk below the surface of the Indian Ocean, which extended along the south of Asia, as it is at present (and probably in direct connection with it), towards the east, as far as further India and the Sunda Islands; towards the west, as far as Madagascar and the south-eastern shores of Africa. We have already mentioned that many facts in animal and vegetable geography render the former existence of such a south Indian continent very probable. (Compare vol. i. p. 361.) Sclater has given this continent the name of Lemuria, from the Semi-apes which were characteristic of it. By assuming this Lemuria to have been man's primæval home, we greatly facilitate the explanation of the geographical distribution of the human species by migration. (Compare the Table of Migrations XV., and its explanation at the end.)

We as yet know of no fossil remains of the hypothetical primæval man (Homo primigenius) who developed out of anthropoid apes during the tertiary period, either in Lemuria or in southern Asia, or possibly in Africa. But considering the extraordinary resemblance between the lowest woolly-haired men, and the highest man-like apes, which still exist at the present day, it requires but a slight stretch of the imagination to conceive an intermediate form connecting the two, and to see in it an approximate likeness to the supposed primæval men, or ape-like men. The form of their skull was probably very long, with slanting teeth; their hair woolly; the colour of their skin dark, of a brownish tint. The hair covering the whole body was probably thicker than in any of the still living human species; their arms comparatively longer and stronger; their legs, on the other hand, knock-kneed, shorter and thinner, with entirely undeveloped calves; their walk but half erect.

This ape-like man very probably did not as yet possess an actual human language, that is, an articulate language of ideas. Human speech, as has already been remarked, most likely originated after the divergence of the primæval species of men into different species. The number of primæval languages is, however, considerably larger than the number of the species of men above discussed. For philologists have hitherto not been able to trace the four primæval languages of the Mediterranean species, namely, the Basque, Caucasian, Semitic, and Indo-Germanic to a single primæval language. As little can the different Negro languages be derived from a common primæval language; hence both these species, Mediterranean and Negro, are certainly polyglottonic, that is, their respective languages originated after the divergence of the speechless primary species into several races had already taken place. Perhaps the Mongols, the Arctic and American tribes, are likewise polyglottonic. The Malayan species is, however, monoglottonic; all the Polynesian and Sundanesian dialects and languages can be derived from a common, long since extinct primæval language, which is not related to any other language on earth. All the other human species, Nubians, Dravidas, Australians, Papuans, Hottentots, and Kaffres are likewise monoglottonic. (Compare p. 333.)

Out of speechless primæval man, whom we consider as the common primary species of all the others, there developed in the first place—probably by natural selection—various species of men unknown to us, and now long since extinct, and who still remained at the stage of speechless ape-men (Alalus, or Pithecanthropus). Two of these species, a woolly-haired and a straight-haired, which were most strongly divergent, and consequently overpowered the others in the struggle for life, became the primary forms of the other remaining human species.

The main branch of woolly-haired men (Ulotrichi) at first spread only over the southern hemisphere, and then emigrated partly eastwards, partly westwards. Remnants of the eastern branch are the Papuans in New Guinea and Melanesia, who in earlier times were diffused much further west (in further India and Sundanesia), and it was not until a late period that they were driven eastwards by the Malays. The Hottentots are the but little changed remnants of the western branch; they immigrated to their present home from the north-east. It was perhaps during this migration that the two nearly related species of Caffres and Negroes branched off from them; but it may be that they owe their origin to a peculiar branch of ape-like men.

The second main branch of primæval straight-haired men (Lissotrichi), which is more capable of development, has probably left a but little changed remnant of its common primary form—which migrated to the south-east—in the ape-like natives of Australia. Probably very closely related to these latter are the South Asiatic primæval Malays, or Promalays, which name we have previously given to the extinct, hypothetical primary form of the other six human species. Out of this unknown common primary form there seem to have arisen three diverging branches, namely, the true Malays, the Mongols, and the Euplocomi; the first spread to the east, the second to the north, and the third westwards.

The primæval home, or the "Centre of Creation," of the Malays must be looked for in the south-eastern part of the Asiatic continent, or possibly in the more extensive continent which existed at the time when further India was directly connected with the Sunda Archipelago and eastern Lemuria. From thence the Malays spread towards the south-east, over the Sunda Archipelago as far as Borneo, then wandered, driving the Papuans before them, eastwards towards the Samoa and Tonga Islands, and thence gradually diffused over the whole of the islands of the southern Pacific, to the Sandwich Islands in the north, the Mangareva in the east, and New Zealand in the south. A single branch of the Malayan tribe was driven far westwards and peopled Madagascar.

The second main branch of primæval Malays, that is, the Mongols, at first also spread in Southern Asia, and, radiating to the east, north, and north-west, gradually peopled the greater part of the Asiatic continent. Of the four principal races of the Mongol species, the Indo-Chinese must perhaps be looked upon as the primary group, out of which at a later period the other Coreo-Japanese and Ural-Altaian races developed as diverging branches. The Mongols migrated in many ways from western Asia into Europe, where the species is still represented in northern Russia and Scandinavia by the Fins and Lapps, in Hungary by the kindred Magyars, and in Turkey by the Osmanlis.

On the other hand, a branch of the Mongols migrated from north-eastern Asia to America, which was probably in earlier times connected with the former continent by a broad isthmus. The Arctic tribes, or Polar men, the Hyper-boreans of north-eastern Asia, and the Esquimaux of the extreme north of America, must probably be regarded as an offshoot of this branch, which became peculiarly degenerated by unfavourable conditions of existence. The principal portion of the Mongolian immigrants, however, migrated to the south, and gradually spread over the whole of America, first over the north, later over South America.

The third and most important main branch of primæval Malays, the curly-haired races, or Euplocomi, have probably left in the Dravidas of Hindostan and Ceylon, that species of man which differs least from the common primary form of the Euplocomi. The principal portion of the latter, namely, the Mediterranean species, migrated from their primæval home (Hindostan?) westwards, and peopled the shores of the Mediterranean, south-western Asia, north Africa, and Europe. The Nubians, in the north-east of Africa, must perhaps be regarded as an offshoot of the primæval Semitic tribes, who migrated far across central Africa almost to the western shores. The various branches of the Indo-Germanic race have deviated furthest from the common primary form of ape-like men. During classic antiquity and the middle ages, the Romanic branch (the Græco-Italo-Keltic group), one of the two main branches of the Indo-Germanic species, outstripped all other branches in the career of civilization, but at present the same position is occupied by the Germanic. Its chief representatives are the English and Germans, who are in the present age laying the foundation for a new period of higher mental development, in the recognition and completion of the theory of descent. The recognition of the theory of development and the monistic philosophy based upon it, forms the best criterion for the degree of man's mental development.

OBJECTIONS AGAINST, AND PROOFS OF THE TRUTH OF, THE THEORY OF DESCENT

If in these chapters I may hope to have made the Theory of Descent seem more or less probable, and to have even convinced some of my readers of its unassailable truth, yet I am by no means unconscious that, to most of them, during the perusal of my explanations, a number of objections more or less well founded must have occurred. Hence it seems absolutely necessary at the conclusion of our examination to refute at least the most important of these, and at the same time, on the other hand, once more to set forth the convincing arguments which bear testimony to the truth of the theory of development.

The objections which are raised to the doctrine of descent may be divided into two large groups: objections of faith and objections of reason. The objections of the first group originate in the infinitely varied forms of faith held by human individuals, and need not here be taken into consideration at all. For, as I have already remarked at the beginning of this book, science, as an objective result of sensuous experience, and of the striving of human reason after knowledge, has nothing whatever to do with the subjective ideas of faith, which are preached by a single man as the direct inspirations or revelations of the Creator, and then believed in by the dependent multitude. This belief, very different in different nations, only begins, as is well known, where science ends. Natural Science believes, according to the maxim of Frederick the Great, "that every one may go to heaven in his own fashion," and only necessarily enters into conflict with particular forms of faith where they appear to set a limit to free inquiry and a goal to human knowledge, beyond which we are not to venture. Now this is certainly the case here in the highest degree, for the Theory of Development applies itself to the solution of the greatest of scientific problems—that of the creation, the coming into existence of things; more especially the origin of organic forms, and of man at their head. It is here certainly the right as well as the sacred duty of free inquiry, to fear no human authority, and courageously to raise the veil from the image of the Creator, unconcerned as to what natural truth may lie concealed beneath. The only Divine revelation which we recognise as true, is written everywhere in nature, and to every one with healthy senses and a healthy reason it is given to participate in the unerring revelation of this holy temple of nature, by his own inquiry and independent discovery.

If we, therefore, here disregard all objections to the Doctrine of Descent which may be raised by the priests of the different religious faiths, we must nevertheless endeavour to refute the most important of those objections which seem more or less founded on science, and which we grant might, at first sight, to a certain extent captivate us and deter us from adopting the Doctrine of Descent. Many persons seem to think the length of the periods of time required the most important of these objections. We are not accustomed to deal with such immense periods as are necessary for the history of the creation. It has already been mentioned that the periods, during which species originated by gradual transmutation, must not be calculated by single centuries, but by hundreds and by millions of centuries. Even the thickness of the stratified crust of the earth, the consideration of the immense space of time which was requisite for its deposition from water, taken together with the periods of elevation between the periods of depression, indicate a duration of time of the organic history of the earth which the human intellect cannot realize. We are here in much the same position as an astronomer in regard to infinite space. In the same way as the distances between the different planetary systems are not calculated by miles but by Sirius-distances, each of which comprises millions of miles, so the organic history of the earth must not be calculated by thousands of years, but by palæontological or geological periods, each of which comprises many thousands of years, and perhaps millions, or even, milliards, of thousands of years. It is of little importance how high the immeasurable length of these periods may be approximately estimated, because we are in fact unable with our limited power of imagination to form a true conception of these periods, and because we do not as in astronomy possess a secure mathematical basis for fixing the approximate length of duration in numbers. But we most positively deny that we see any objection to the theory of development in the extreme length of these periods which are so completely beyond the power of our imagination. It is, on the contrary, as I have already explained in one of the preceding chapters, most advisable, from a strictly philosophical point of view, to conceive these periods of creation to be as long as possible, and we are by so much the less in danger of losing ourselves in improbable hypotheses, the longer we conceive the periods for organic processes of development to have been. The longer, for example, we conceive the Permian period to have been, the easier it will be for us to understand how the important transmutations took place within it which so essentially distinguish the fauna and flora of the Coal period from that of the Trias. The great disinclination which most persons have to assume such immeasurable periods, arises mainly from the fact of our having in early youth been brought up in the notion that the whole earth is only some thousands of years old. Moreover, human life, which at most attains the length of a century, is an extremely short space of time, and is not suitable as a standard for the measurement of geological periods. Our life is a single drop in the ocean of eternity. The reader may call to mind the duration of life of many trees which is more than fifty times as long; for example, the dragon-trees (Dracæna) and monkey bread-fruit trees (Adansonia), whose individual life exceeds a period of five thousand years; and, on the other hand, the shortness of the individual life of many of the lower animals, for example, the infusoria, where the individual, as such, lives but a few days, or even but a few hours, contrasts no less strongly with human longevity. This comparison brings the relative nature of all measurement of time very clearly before us. If the theory of development be true at all, there must certainly have elapsed immense periods, utterly inconceivable to us, during which the gradual historical development of the animal and vegetable kingdom proceeded by the slow transformation of species. There is, however, not a single reason for accepting a definite limit for the length of these periods of development.

A second main objection which many, and more especially systematic zoologists and botanists, raise against the theory of descent, is that no transition forms between the different species can be found, although according to the theory of descent they ought to be found in great numbers. This objection is partly well founded and partly not so, for there does exist an extraordinarily large number of transition forms between living, as well as between extinct species, especially where we have an opportunity of seeing and comparing very numerous individuals of kindred species. Those careful investigators of individual species who so frequently raise this objection are the very persons whom we constantly find checked in their special series of investigations by the really insuperable difficulty of sharply distinguishing individual species. In all systematic works, which are in any degree thorough, one meets with endless complaints, that here and there species cannot be distinguished because of the excessive number of transition forms. Hence every naturalist defines the limit and the number of individual species differently. Some zoologists and botanists, as I mentioned (vol. i. p. 276), assume in one and the same group of organisms ten species, others twenty, others a hundred or more, while other systematic naturalists again look upon these different forms only as varieties of a single "good" species. In most groups of forms there is, in fact, a superabundance of transition forms and intermediate stages between the individual species.

It is true that in many species the forms of transition are actually wanting, but this is easily explained by the principle of divergence or separation, the importance of which I have already explained. The circumstance that the struggle for existence is the more active between two kindred forms the closer they stand to each other, must necessarily favour the speedy extinction of the connecting intermediate forms between the two divergent species. If one and the same species produce diverging varieties in different directions, which become new species, the struggle between these new forms and the common primary form will be the keener the less they differ from one another; but the stronger the divergence the less dangerous the struggle. Naturally therefore, it is principally the connecting intermediate forms which will in most cases quietly die out, while the most divergent forms remain and reproduce themselves as distinct "new species." In accordance with this, we in fact no longer find forms of transition leading to those groups which are becoming extinct, as, for example, among birds, are the ostriches; and among mammals, the elephants, giraffes, Semi-apes, Edentata, and Ornithorhyncus. The groups of forms approaching their extinction no longer produce new varieties, and naturally the species are what is called "good," that is, the species are distinctly different from one another. But in those animal groups where development and progress are still active, where the existing species deviate into many new species by the formation of new varieties, we find an abundance of transition forms which cause the greatest difficulties to systematic naturalists. This is the case, for example, among birds with the finches; among mammals with most of the rodents (more especially with those of the mouse and rat kind), with a number of the ruminants and with genuine apes, more especially with the South American forms (Cebus), and many others. The continual development of species by the formation of new varieties here produces a mass of intermediate forms which connect the so-called "good" species, which efface their boundaries, and render their sharp specific distinction completely illusory.

The reason that this nevertheless does not cause a complete confusion of forms, nor a universal chaos in the structure of animals and vegetables, lies simply in the fact that there is a continual counteraction at work between progressive adaptation on the one hand, and the retentive power of inheritance on the other hand. The degree of stability and variability manifested by every organic form is determined solely by the actual condition of the equilibrium between these two opposite functions. Inheritance is the cause of the stability of species, adaptation the cause of their modification. When therefore some naturalists say that, according to the theory of descent, there ought to be a much greater variety of forms, and others again, that there ought to be a much greater equality of forms, the former under-estimate the value of inheritance and the latter the value of adaptation. The ratio of the interaction between inheritance and adaptation determines the ratio of the stability and variability of organic species at any given period.

Another objection to the theory of descent, which, in the opinion of many naturalists and philosophers is of great weight, is that it ascribes the origin of organs which act for a definite purpose to causes which are either aimless or mechanical in their operation. This objection seems to be especially important in regard to those organs which appear so excellently adapted for a certain definite purpose that the most ingenious mechanician could not invent a more perfect organ for the purpose. Such are, above all, the higher sense-organs of animals, the eye and ear. If the eyes and auditory apparatus of the higher animals alone were known to us, they would indeed cause great and perhaps insurmountable difficulties. How could we come to the conclusion that the extraordinarily great and wonderful degree of perfection and conformity to purpose which we perceive in the eyes and ears of higher animals, is in every respect attained solely by natural selection? Fortunately, however, comparative anatomy and the history of development help us here over all obstacles; for when in the animal kingdom we follow the gradual progress towards perfection of the eyes and ears, step by step, we find such a finely graduated series of improvement, that we can clearly follow the development of the most complex organs through all the stages towards perfection. Thus, for example, the eye in the lowest animal is a simple spot of pigment which does not yet reflect any image of external objects, but at most perceives and distinguishes the different rays of light. Later, we find in addition to this a sensitive nerve; then there gradually develops within the spot of pigment the first beginning of the lens, a refractive body which is now able to concentrate the rays of light and to reflect a definite image. But all the composite apparatus for the movement of the eye and its accommodation to variations of light and distance are still absent, namely, the various refractive media, the highly differentiated membrane of the optic nerve, etc., which are so perfectly constructed in higher animals. Comparative anatomy shows us an uninterrupted succession of all possible stages of transition, from the simplest organ to the most highly perfected apparatus, so that we can form a pretty correct idea of the slow and gradual formation of even such an exceedingly complex organ. The like gradual progress which we observe in the development of the organ during the course of individual development, must have taken place in the historical (phyletic) origin of the organ.

Many persons when contemplating these most perfect organs—which apparently were purposely invented and constructed by an ingenious Creator for a definite function, but which in reality have arisen by the aimless action of natural selection—experience difficulties in arriving at a rational understanding of them, which are similar to those experienced by the uncivilized tribes of nature when contemplating the latest complicated productions of engineering. Savages who see a ship of the line, or a locomotive engine for the first time, look upon these objects as the productions of a supernatural being, and cannot understand how a man, an organism like themselves, could have produced such an engine. Even the uneducated classes of our own race cannot comprehend such an intricate apparatus in its actual workings, nor can they understand its purely mechanical nature. Most naturalists, however, as Darwin very justly remarks, stand in much the same position in regard to the forms of organisms as do savages to ships of the line and to locomotive engines. A rational understanding of the purely mechanical origin of organic forms can only be acquired by a thorough and general training in Biology, and by a special knowledge of comparative anatomy and the history of development. Among the remaining objections to the Theory of Descent, I shall here finally refer to and refute but one more, as in the eyes of many unscientific men it seems to possess great weight. How are we, from the Theory of Descent, to conceive of the origin of the mental faculties of animals, and more especially their specific expressions—the so-called instincts? This difficult subject has been so minutely discussed by Darwin in a special chapter of his chief work (the seventh), that I must refer the reader to it. We must regard instincts as essentially the habits of the soul acquired by adaptation, and transmitted and fixed by inheritance through many generations. Instincts are, therefore, like all other habits, which, according to the laws of cumulative adaptation (vol. i. p. 233) and established inheritance (vol. i. p. 216), lead to the origin of new functions, and thus also to new forms of the organs. Here, as everywhere, the interaction between function and organ goes hand in hand. Just as the mental faculties of man have been acquired by the progressive adaptation of the brain, and been fixed by continual transmission by inheritance, so the instincts of animals, which differ from them only in quantity, not in quality, have arisen by the gradual perfecting of their mental organ, that is, their central nervous system, by the interaction of Adaptation and Inheritance. Instincts, as is well known, are inherited, but experiences and, consequently, new adaptations of the animal mind, are also transmitted by inheritance; and the training of domestic animals to different mental activities, which wild animals are incapable of accomplishing, rests upon the possibility of mental adaptation. We already know a series of examples, in which such adaptations, after they had been transmitted through a succession of generations, finally appeared as innate instincts, and yet they have only been acquired from the ancestors of the animals. Inheritance has here caused the result of training to become instinct. The characteristic instincts of sporting dogs, shepherd's dogs, and other domestic animals, and the natural instincts of wild animals, which they possess at birth, were in the first place acquired by their ancestors by adaptation. They may in this respect be compared to man's "knowledge a priori," which, like all other knowledge, was originally acquired by our remote ancestors, "a posteriori," by sensuous experience. As I have already remarked, it is evident that "knowledge a priori" arose only by long-enduring transmission, by inheritance of acquired adaptations of the brain, out of originally empiric or experiential "knowledge a posteriori" (vol. i. p. 31).

The objections to the Theory of Descent here discussed and refuted are, I believe, the most important which have been raised against it; I consider also that I have sufficiently proved to the reader their futility. The numerous other objections which besides these have been raised against the Theory of Development in general, or against its biological part, the Theory of Descent in particular, arise either from such a degree of ignorance of empirically established facts, or from such a want of their right understanding, and from such an incapacity to draw the necessary conclusions, that it is really not worth the trouble to go further into the refutation. There are only some general points in regard to which, I should like, in a few words, to draw attention.

In the first place I must observe, that in order thoroughly to understand the doctrine of descent, and to be convinced of its absolute truth, it is indispensable to possess a general knowledge of the whole of the domain of biological phenomena. The theory of descent is a biological theory, and hence it may with fairness and justice be demanded that those persons who wish to pass a valid judgment upon it should possess the requisite degree of biological knowledge. Their possessing a special empiric knowledge of this or that domain of zoology or botany, is not sufficient; they must possess a general insight into the whole series of phenomena, at least in the case of one of the three organic kingdoms. They ought to know what universal laws result from the comparative morphology and physiology of organisms, but more especially from comparative anatomy, from the individual and the palæontological history of development, etc.; and they ought to have some idea of the deep mechanical, causal connection between all these series of phenomena. It is self-evident that a certain degree of general culture, and especially a philosophical education, is requisite; which is, however, unfortunately by many persons in our day, not considered at all necessary. Without the necessary connection of empirical knowledge and the philosophical understanding of biological phenomena, it is impossible to gain a thorough conviction of the truth of the Theory of Descent.

Now I ask, in the face of this first preliminary condition for a true understanding of the Theory of Descent, what we are to think of the confused mass of persons who have presumed to pass a written or oral judgment upon it of an adverse character? Most of them are unscientific persons, who either know nothing of the most important phenomena of Biology, or at least possess no idea of their deeper significance. What should we say of an unscientific person who presumed to express an opinion on the cell-theory, without ever having seen cells; or of one who presumed to question the vertebral-theory, without ever having studied comparative anatomy? And yet one may meet with such ridiculous arrogance any day in the history of the biological Theory of Descent. One hears thousands of unscientific and but half-educated persons pass a final judgment upon it, although they know nothing either of botany or of zoology, of comparative anatomy or the theory of tissues, of palæontology or embryology. Hence it happens, as Huxley well says, that most of the writings published against Darwin are not worth the paper upon which they are written.

We might add that there are many naturalists, and even celebrated zoologists and botanists, among the opponents of the Theory of Descent; but these latter are mostly old stagers, who have grown grey in quite opposite views, and whom we cannot expect, in the evening of their lives, to submit to a reform in their conception of the universe, which has become to them a fixed idea.

It is, moreover, expressly to be remarked, that not only a general insight into the whole domain of biological phenomena, but also a philosophical understanding of it, are the necessary preliminary conditions for becoming convinced of and adopting the Theory of Descent. Now we shall find that these indispensable preliminary conditions are, unfortunately, by no means fulfilled by the majority of naturalists of the present day. The immense amount of empirical facts with which the gigantic advances of modern natural science have recently made us acquainted has led to a prevailing inclination for the special study of single phenomena and of small and narrow domains. This causes the knowledge of other paths, and especially of Nature as a great comprehensive whole, to be in most cases completely neglected. Every one with sound eyes and a microscope, together with industry and patience for study, can in our day attain a certain degree of celebrity by microscopic "discoveries," without, however, deserving the name of a naturalist. This name is deserved only by him who not merely strives to know the individual phenomena, but who also seeks to discover their causal connection. Even in our own day, most palæontologists examine and describe fossils without knowing the most important facts of embryology. Embryologists, on the other hand, follow the history of development of a particular organic individual, without having an idea of the palæontological history of the whole tribe, of which fossils are the records. And yet these two branches of the organic history of development—ontogeny, or the history of the individual, and phylogeny, or the history of the tribe—stand in the closest causal connection, and the one cannot be understood without the other. The same may be said of the systematic and the anatomical part of Biology. There are even now, in zoology and botany, many systematic naturalists who work with the erroneous idea that it is possible to construct a natural system of animals and plants simply by a careful examination of the external and readily accessible forms of bodies, without a deeper knowledge of their internal structure. On the other hand, there are anatomists and histologists who think it possible to obtain a true knowledge of animal and vegetable bodies merely by a most careful examination of the inner structure of the body of some individual species, without the comparative examination of the bodily form of all kindred organisms. And yet here, as everywhere, the internal and external factors, to wit, Inheritance and Adaptation, stand in the closest mutual relation, and the individual can never be thoroughly understood without a comparison of it with the whole of which it is a part. To those one-sided specialists we should like in Goethe's words to say:—

We must, contemplating Nature,Part as Whole, give equal heed to:Nought is inward, nought is outward,For the inner is the outer.6

And again:—

Nature has neither kernel nor shell,It is she that is All and All at once.7

What is even more detrimental to the general understanding of nature as a whole than this one-sided tendency, is the want of a philosophical culture, and this applies to most of the naturalists of the present day. The various errors of the earlier speculative nature-philosophy made during the first thirty years of our century, have brought the whole of philosophy into such bad repute with the exact empirical naturalists, that they live in the strange delusion that it is possible to erect the edifice of natural science out of mere facts, without their philosophic connection; in short, out of mere knowledge, without the understanding of it. But as a purely speculative and absolutely philosophical system, which does not concern itself with the indispensable foundation of empirical facts, becomes a castle in the air, which the first real experiment throws to the winds; so, on the other hand, a purely empirical system, constructed of nothing but facts, remains a disorderly heap of stones, which will never deserve the name of an edifice. Bare facts established by experience are nothing but rude stones, and without their thoughtful valuation, without their philosophic connection, no science can be established. As I have already tried to impress upon my reader, the strong edifice of true monistic science, or what is the same thing, the Science of Nature, exists only by the closest interaction, and the reciprocal penetration of philosophy and empirical knowledge.

This lamentable estrangement between science and philosophy, and the rude empiricism which is now-a-days unfortunately praised by most naturalists as "exact science," have given rise to those strange freaks of the understanding, to those gross insults against elementary logic, and to that incapacity for forming the simplest conclusions which one may meet with any day in all branches of science, but especially in zoology and botany. It is here that the neglect of a philosophical culture and training of the mind, directly avenges itself most painfully. It is not to be wondered at that the deep inner truth of the Theory of Descent remains a sealed book to those rude empiricists. As the common proverb justly says: they cannot see the wood for the trees. It is only by a more general philosophical study, and especially by a more strictly logical training of the mind, that this sad state of things can be remedied. (Compare Gen. Morph. i. 63; ii. p. 447.)

If we rightly consider this circumstance, and if we further reflect upon it in connection with the empirical foundation of the philosophical theory of development, we shall at once see how we are placed respecting the oft-demanded proofs of the theory of descent. The more the doctrine of filiation has of late years made way for itself, and the more all thoughtful, younger naturalists, and all truly biologically-educated philosophers have become convinced of its inner truth and absolute necessity, the louder have its opponents called for actual proofs. The same persons who, shortly after the publication of Darwin's work, declared it to be "a groundless, fantastic system," an "arbitrary speculation," an "ingenious dream," now kindly condescend to declare that the theory of descent certainly is a scientific "hypothesis," but that it still requires to be "proved." When these remarks are made by persons who do not possess the requisite empirico-philosophical culture, nor the necessary knowledge in comparative anatomy, embryology, and palæontology, we cannot be much offended, and we refer them to the study of those sciences. But when similar remarks are made by acknowledged specialists, by teachers of zoology and botany, who certainly ought to possess a general insight into the whole domain of their science, or who are actually familiar with the facts of those scientific domains, then we are really at a loss what to say. Those who are not satisfied with the treasures of our present empirical knowledge of nature as a basis on which to establish the Theory of Descent, will not be convinced by any other facts which may hereafter be discovered; for we can conceive no circumstances which would furnish stronger or a more complete testimony to the truth of the doctrine of filiation than is even now seen, for example, in the well-known facts of comparative anatomy and ontogeny. I must here again direct attention to the fact, that all the great and general laws, and all the comprehensive series of phenomena of the most different domains of biology can only be explained and understood by the Theory of Development (and especially by its biological part, the Theory of Descent), and that without it they remain completely inexplicable and incomprehensible. The internal causal connection between them all proves the Theory of Descent to be the greatest inductive law of Biology.

Before concluding, I will once more name all those series of inductions, all those general laws of Biology, upon which this comprehensive law of development is firmly based.

(1.) The palæontological history of the development of organisms, the gradual appearance and the historical succession of the different species and groups of species, the empirical laws of the palæontological change of species, as furnished to us by the science of fossils, and more especially the progressive differentiation and perfecting of animal and vegetable groups in the successive periods of the earth's history.

(2.) The individual history of development of organisms, embryology and metamorphology, the gradual changes in the slow development of the body and its particular organs, especially the progressive differentiation and perfecting of the organs and parts of the body in the successive periods of the individual development.

(3.) The inner causal connection between ontogeny and phylogeny, the parallelism between the individual history of the development of organisms, and the palæontological history of the development of their ancestors, a connection which is actually established by the laws of Inheritance and Adaptation, and which may be summed up in the words: ontogeny, according to the laws of inheritance and adaptation, repeats in its large features the outlines of phylogeny.

(4.) The comparative anatomy of organisms, the proof of the essential agreement of the inner structure of kindred organisms, in spite even of the greatest difference of external form in the various species; their explanation by the causal dependence of the internal agreement of the structure on Inheritance, the external dissimilarity of the bodily form on Adaptation.

(5.) The inner causal connection between comparative anatomy and the history of development, the harmonious agreement between the laws of the gradual development, the progressive differentiation and perfecting, as they may be seen in comparative anatomy on the one hand, in ontogeny and palæontology on the other.

(6.) Dysteleology, or the theory of purposelessness, the name I have given to the science of rudimentary organs, of suppressed and degenerated, aimless and inactive, parts of the body; one of the most important and most interesting branches of comparative anatomy, which, when rightly estimated, is alone sufficient to refute the fundamental error of the teleological and dualistic conception of Nature, and to serve as the foundation of the mechanical and monistic conception of the universe.

(7.) The natural system of organisms, the natural grouping of all the different forms of Animals, Plants, and Protista into numerous smaller or larger groups, arranged beside and above one another; the kindred connection of species, genera, families, orders, classes, tribes, etc., more especially, however, the arboriform branching character of the natural system, which is the spontaneous result of a natural arrangement and classification of all these graduated groups or categories. The result attained in attempting to exhibit the relationships of the mere forms of organisms by a tabular classification is only explicable when regarded as the expression of their actual blood relationship; the tree shape of the natural system can only be understood as the actual pedigree of the organisms.

(8.) The chorology of organisms, the science of the local distribution of organic species, of their geographical and topographical dispersion over the surface of the earth, over the heights of mountains and in the depths of the ocean, but especially the important phenomenon that every species of organism proceeds from a so-called "centre of creation" (more correctly a "primæval home" or "centre of distribution"); that is, from a single locality, where it originated but once, and whence it spread.

(9.) The œcology of organisms, the knowledge of the sum of the relations of organisms to the surrounding outer world, to organic and inorganic conditions of existence; the so-called "economy of nature," the correlations between all organisms living together in one and the same locality, their adaptation to their surroundings, their modification in the struggle for existence, especially the circumstances of parasitism, etc. It is just these phenomena in "the economy of nature" which the unscientific, on a superficial consideration, are wont to regard as the wise arrangements of a Creator acting for a definite purpose, but which on a more attentive examination show themselves to be the necessary results of mechanical causes.

(10.) The unity of Biology as a whole, the deep inner connection existing between all the phenomena named and all the other phenomena belonging to zoology, protistics, and botany, and which are simply and naturally explained by a single common principle. This principle can be no other than the common derivation of all the specifically different organisms from a single, or from several absolutely simple, primary forms like the Monera, which possess no organs. The Theory of Descent, by assuming this common derivation, throws a clear light upon these individual series of phenomena, as well as upon their totality, without which their deeper causal connection would remain completely incomprehensible to us. The opponents of the Theory of Descent can in no way explain any single one of these series of phenomena or their deeper connection with one another. So long as they are unable to do this, the Theory of Descent remains the one adequate biological theory.

We should, on account of the grand proofs just enumerated, have to adopt Lamarck's Theory of Descent for the explanation of biological phenomena, even if we did not possess Darwin's Theory of Selection. The one is so completely and directly proved by the other, and established by mechanical causes, that there remains nothing to be desired. The laws of Inheritance and Adaptation are universally acknowledged physiological facts, the former traceable to propagation, the latter to the nutrition of organisms. On the other hand, the struggle for existence is a biological fact, which with mathematical necessity follows from the general disproportion between the average number of organic individuals and the numerical excess of their germs. But as Adaptation and Inheritance in the struggle for life are in continual interaction, it inevitably follows that natural selection, which everywhere influences and continually changes organic species, must, by making use of divergence of character, produce new species. Its influence is further especially favoured by the active and passive migrations of organisms, which go on everywhere. If we give these circumstances due consideration, the continual and gradual modification or transmutation of organic species will appear as a biological process, which must, according to causal law, of necessity follow from the actual nature of organisms and their mutual correlations.

That even the origin of man must be explained by this general organic process of transmutation, and that it is simply as well as naturally explained by it, has, I believe, been sufficiently proved in my last chapter but one. I cannot, however, avoid here once more directing attention to the inseparable connection between this so-called "theory of apes," or "pithecoid theory," and the whole Theory of Descent. If the latter is the greatest inductive law of biology, then it of necessity follows that the former is its most important deductive law. They stand and fall together. As all depends upon a right understanding of this proposition, which in my opinion is very important, and which I have therefore several times brought before the reader, I may be allowed to explain it here by an example.

In all mammals known to us the centre of the nervous system is the spinal marrow and the brain, and the centre of the vascular system is a quadrupal heart, consisting of two principal chambers and two ante-chambers. From this we draw the general inductive conclusion that all mammals, without exception, those extinct, together with all those living species as yet unknown to us, as well as the species which we have examined, possess a like organization, a like heart, brain, and spinal marrow. Now if, as still happens every year, there be discovered in any part of the earth a new species of mammal, a new species of marsupial, or a new species of deer, or a new species of ape, every zoologist knows with certainty at once, without having examined its inner structure, that this species must possess a quadruple heart, a brain and spinal marrow, like all other mammals. Not a single naturalist would ever think of supposing that the central nervous system of this new species of mammal could possibly consist of a ventral cord with an œsophageal collar as in the insects, or of scattered pairs of knots as in the molluscs, or that its heart could be many-chambered as in flies, or one-chambered as in the tunicates. This completely certain and safe conclusion, although it is not based upon any direct experience, is a deductive conclusion. In the same way, as I have shown in a previous chapter, Goethe, from the comparative anatomy of mammals, established the general inductive conclusion that they all possess a mid jawbone, and afterwards drew from it the special deductive conclusion that man, who in all other respects does not essentially differ from other mammals, must also possess a like mid jawbone. He maintained this conclusion without having actually seen the human mid jawbone, and only proved its existence subsequently by actual observation (vol. i. p. 84).

The process of induction is a logical system of forming conclusions from the special to the general, by which we advance from many individual experiences to a general law; deduction, on the other hand, draws a conclusion from the general to the special, from a general law of nature to an individual case. Thus the Theory of Descent is, without doubt, a great inductive law, empirically based upon all the biological experience cited above; the pithecoid theory, on the other hand, which asserts that man has developed out of lower, and in the first place out of ape-like mammals, is a deductive law inseparably connected with the general inductive law.

The pedigree of the human race, the approximate outlines of which I gave in the last chapter but one, of course remains in detail (like all the pedigrees of animals and plants previously discussed) a more or less approximate general hypothesis. This however does not affect the application of the theory of descent to man. Here, as in all investigations on the derivation of organisms, one must clearly distinguish between the general theory of descent and the special hypotheses of descent. The general theory of descent claims full and lasting value, because it is an inductive law, based upon all the whole series of biological phenomena and their inner causal connection. Every special hypothesis of descent, on the other hand, has its special value determined by the existing condition of our biological knowledge, and by the extent of the objective empirical basis upon which we deductively establish this particular hypothesis. Hence, all the individual attempts to obtain a knowledge of the pedigree of any one group of organisms possesses but a temporary and conditional value, and any special hypothesis relating to it will become the more and more perfect the greater the advance we make in the comparative anatomy, ontogeny, and palæontology of the group in question. The more, however, we enter into genealogical details, and the further we trace the separate off-shoots and branches of the pedigree, the more uncertain and subjective becomes our special hypothesis of descent on account of the incompleteness of our empirical basis. This however does no injury to the general theory of descent, which remains as the indispensable foundation for really profound apprehension of biological phenomena. Accordingly, there can be no doubt that we can and must, with full assurance, regard the derivation of man—in the first place, from ape-like forms; farther back, from lower mammals, and thus continually farther back to lower stages of the vertebrata down to their lowest invertebrate roots, nay, even down to a simple plastid—as a general theory. On the other hand, the special tracing of the human pedigree, the closer definition of the animal forms known to us, which either actually belong to the ancestors of man, or at least stand in very close blood relationship to them, will always remain a more or less approximate hypothesis of descent, all the more in danger of deviating from the real pedigree the nearer it endeavours to approach it by searching for the individual ancestral forms. This state of things results from the immense gaps in our palæontological knowledge, which can, under no circumstances, ever attain to even an approximate completeness.

A thoughtful consideration of this important circumstance at once furnishes the answer to a question which is commonly raised in discussing this subject, namely, the question of scientific proofs for the animal origin of the human race. Not only the opponents of the Theory of Descent, but even many of its adherents who are wanting in the requisite philosophical culture, look too much for "signs" and for special empirical advances in the science of nature. They await the sudden discovery of a human race with tails, or of a talking species of ape, or of other living or fossil transition forms between man and the ape, which shall fill the already narrow chasm between the two, and thus empirically "prove" the derivation of man from apes. Such special manifestations, were they ever so convincing and conclusive, would not furnish the proof desired. Unthinking persons, or those unacquainted with the series of biological phenomena, would still be able to maintain the objections to those special testimonies which they now maintain against our theory.

The absolute certainty of the Theory of Descent, even in its application to man, is built on a more solid foundation; and its true inner value can never be tested simply by reference to individual experience, but only by a philosophical comparison and estimation of the treasures of all our biological experiences. The inestimable importance of the Theory of Descent is surely based upon this, that the theory follows of necessity (as a general inductive law) from the comparative synthesis of all organic phenomena of nature, and more especially from the triple parallelism of comparative anatomy, of ontogeny, and phylogeny; and the pithecoid theory under all circumstances (apart from all special proofs) remains as a special deductive conclusion which must of necessity be drawn from the general inductive law of the Theory of Descent.

In my opinion, all depends upon a right understanding of this philosophical foundation of the Theory of Descent and of the pithecoid theory which is inseparable from it. Many persons will probably admit this, and yet at the same time maintain that all this applies only to the bodily, not to the mental development of man. Now, as we have hitherto been occupied only with the former, it is perhaps necessary here to cast a glance at the latter, in order to show that it is also subject to the great general law of development. In doing this it is above all necessary to recollect that body and mind can in fact never be considered as distinct, but rather that both sides of nature are inseparably connected, and stand in the closest interaction. As even Goethe has clearly expressed it—"matter can never exist and act without mind, and mind never without matter." The artificial discord between mind and body, between force and matter, which was maintained by the erroneous dualistic and teleological philosophy of past times has been disposed of by the advances of natural science, and especially by the theory of development, and can no longer exist in face of the prevailing mechanical and monistic philosophy of our day. How human nature, and its position in regard to the rest of the universe, is to be conceived of according to the modern view, has been minutely discussed by Radenhausen in his "Isis,"(33) which is excellent and well worth perusal.

With regard to the origin of the human mind or the soul of man, we, in the first place, perceive that in every human individual it develops from the beginning, step by step and gradually, just like the body. In a newly born child we see that it possesses neither an independent consciousness, nor in fact clear ideas. These arise only gradually when, by means of sensuous experience, the phenomena of the outer world affect the central nervous system. But still the little child is wanting in all those differentiated emotions of the soul which the full-grown man acquires only by the long experience of years. From this graduated development of the human soul in every single individual we can, in accordance with the inner causal connection between ontogeny and phylogeny, directly infer the gradual development of the human soul in all mankind, and further, in the whole of the vertebrate tribe. In its inseparable connection with the body, the human soul or mind has also had to pass through all those gradual stages of development, all those various degrees of differentiation and perfecting, of which the hypothetical series of human ancestors sketched in a late chapter gives an approximate representation.

It is true that this conception generally greatly offends most persons on their first becoming acquainted with the Theory of Development, because more than all others it most strongly contradicts the traditional and mythological ideas, and the prejudices which have been held sacred for thousands of years. But like all other functions of organisms, the human soul must necessarily have historically developed, and the comparative or empirical study of animal psychology clearly shows that this development can only be conceived of as a gradual evolution from the soul of vertebrate animals, as a gradual differentiation and perfecting which, in the course of many thousands of years, has led to the glorious triumph of the human mind over its lower animal ancestral stages. Here, as everywhere, the only way to arrive at a knowledge of natural truth is to compare kindred phenomena, and investigate their development. Hence we must above all, as we did in the examination of the bodily development, compare the highest animal phenomena on the one hand with the lowest animal phenomena, and on the other with the lowest human phenomena. The final result of this comparison is this—that between the most highly developed animal souls, and the lowest developed human souls, there exists only a small quantitative, but no qualitative difference, and that this difference is much less than the difference between the lowest and the highest human souls, or than the difference between the highest and the lowest animal souls.

In order to be convinced of this important result, it is above all things necessary to study and compare the mental life of wild savages and of children.(32) At the lowest stage of human mental development are the Australians, some tribes of the Polynesians, and the Bushmen, Hottentots, and some of the Negro tribes. Language, the chief characteristic of genuine men, has with them remained at the lowest stage of development, and hence also their formation of ideas has remained at a low stage. Many of these wild tribes have not even a name for animal, plant, colour, and such most simple ideas, whereas they have a word for every single, striking form of animal and plant, and for every single sound or colour. Thus even the most simple abstractions are wanting. In many of these languages there are numerals only for one, two, and three: no Australian language counts beyond four. Very many wild tribes can count no further than ten or twenty, whereas some very clever dogs have been made to count up to forty and even beyond sixty. And yet the faculty of appreciating number is the beginning of mathematics! Nothing, however, is perhaps more remarkable in this respect, than that some of the wildest tribes in southern Asia and eastern Africa have no trace whatever of the first foundations of all human civilization, of family life, and marriage. They live together in herds, like apes, generally climbing on trees and eating fruits; they do not know of fire, and use stones and clubs as weapons, just like the higher apes. All attempts to introduce civilization among these, and many of the other tribes of the lowest human species, have hitherto been of no avail; it is impossible to implant human culture where the requisite soil, namely, the perfecting of the brain, is wanting. Not one of these tribes has ever been ennobled by civilization; it rather accelerates their extinction. They have barely risen above the lowest stage of transition from man-like apes to ape-like men, a stage which the progenitors of the higher human species had already passed through thousands of years ago.(44)

Now consider, on the other hand, the highest stages of development of mental life in the higher vertebrate animals, especially birds and mammals. If, as is usually done, we divide the different emotions of the soul into three principal groups—sensation, will, and thought—we shall find in regard to every one of them, that the most highly developed birds and mammals are on a level with the lowest human beings, or even decidedly surpass them. The will is as distinctly and strongly developed in higher animals as in men of character. In both cases it is never actually free, but always determined by a causal chain of ideas. (Compare vol. i. p. 237.) In like manner, the different degrees of will, energy, and passion are as variously graduated in higher animals as in man. The affections of the higher animals are not less tender and warm than those of man. The fidelity and devotion of the dog, the maternal love of the lioness, the conjugal love and connubial fidelity of doves and love-birds are proverbial, and might serve as examples to many men. If these virtues are to be called "instincts," then they deserve the same name in mankind. Lastly, with regard to thought, the comparative consideration of which doubtless presents the most difficulties, this much may with certainty be inferred—especially from an examination of the comparative psychology of cultivated domestic animals—that the processes of thinking, here follow the same laws as in ourselves. Experiences everywhere form the foundation of conceptions, and lead to the recognition of the connection between cause and effect. In all cases, as in man, it is the path of induction and deduction which leads to the formation of conclusions. It is evident that in all these respects the most highly developed animals stand much nearer to man than to the lower animals, although they are also connected with the latter by a chain of gradual and intermediate stages. In Wundt's excellent "Lectures on the Human and Animal Soul,"(46) there are a number of proofs of this.

Now, if instituting comparisons in both directions, we place the lowest and most ape-like men (the Austral Negroes, Bushmen, and Andamans, etc.), on the one hand, together with the most highly developed animals, for instance, with apes, dogs, and elephants, and on the other hand, with the most highly developed men—Aristotle, Newton, Spinoza, Kant, Lamarck, or Goethe—we can then no longer consider the assertion, that the mental life of the higher mammals has gradually developed up to that of man, as in any way exaggerated. If one must draw a sharp boundary between them, it has to be drawn between the most highly developed and civilized man on the one hand, and the rudest savages on the other, and the latter have to be classed with the animals. This is, in fact, the opinion of many travellers, who have long watched the lowest human races in their native countries. Thus, for example, a great English traveller, who lived for a considerable time on the west coast of Africa, says: "I consider the negro to be a lower species of man, and cannot make up my mind to look upon him as 'a man and a brother,' for the gorilla would then also have to be admitted into the family." Even many Christian missionaries, who, after long years of fruitless endeavours to civilize these lowest races, have abandoned the attempt, express the same harsh judgment, and maintain that it would be easier to train the most intelligent domestic animals to a moral and civilized life, than these unreasoning brute-like men. For instance, the able Austrian missionary Morlang, who tried for many years without the slightest success to civilize the ape-like negro tribes on the Upper Nile, expressly says: "that any mission to such savages is absolutely useless. They stand far below unreasoning animals; the latter at least show signs of affection towards those who are kind towards them, whereas these brutal natives are utterly incapable of any feeling of gratitude."

Now, it clearly follows from these and other testimonies, that the mental differences between the lowest men and the animals are less than those between the lowest and the highest men; and if, together with this, we take into consideration the fact that in every single human child mental life develops slowly, gradually, and step by step, from the lowest condition of animal unconsciousness, need we still feel offended when told that the mind of the whole human race has in like manner gone through a process of slow, gradual, and historical development? Can we find it "degrading" to the human soul that, by a long and slow process of differentiation and perfecting, it has very gradually developed out of the soul of vertebrate animals? I freely acknowledge that this objection, which is at present raised by many against the pithecoid theory, is quite incomprehensible to me. On this point Bernhard Cotta, in his excellent "Geologie der Gegenwart," very justly remarks: "Our ancestors may be a great honour to us; but it is much better if we are an honour to them!"(31)367

Our Theory of Development explains the origin of man and the course of his historical development in the only natural manner. We see in his gradually ascensive development out of the lower vertebrata, the greatest triumph of humanity over the whole of the rest of Nature. We are proud of having so immensely outstripped our lower animal ancestors, and derive from it the consoling assurance that in future also, mankind, as a whole, will follow the glorious career of progressive development, and attain a still higher degree of mental perfection. When viewed in this light, the Theory of Descent as applied to man opens up the most encouraging prospects for the future, and frees us from all those anxious fears which have been the scarecrows of our opponents.

We can even now foresee with certainty that the complete victory of our Theory of Development will bear immensely rich fruits—fruits which have no equal in the whole history of the civilization of mankind. Its first and most direct result—the complete reform of Biology—will necessarily be followed by a still more important and fruitful reform of Anthropology.

From this new theory of man there will be developed a new philosophy, not like most of the airy systems of metaphysical speculation hitherto prevalent, but one founded upon the solid ground of Comparative Zoology. A beginning of this has already been made by the great English philosopher Herbert Spencer.(45) Just as this new monistic philosophy first opens up to us a true understanding of the real universe, so its application to practical human life must open up a new road towards moral perfection. By its aid we shall at last begin to raise ourselves out of the state of social barbarism in which, notwithstanding the much vaunted civilization of our century, we are still plunged. For, unfortunately, it is only too true, as Alfred Wallace remarks with regard to this, at the end of his book of travels: "Compared with our wondrous progress in physical science and its practical applications, our system of government, of administering justice, of national education, and our whole social and moral organisation remains in a state of barbarism."

This social and moral barbarism we shall never overcome by the artificial and perverse training, the one-sided and defective teaching, the inner untruth and the external tinsel, of our present state of civilization. It is above all things necessary to make a complete and honest return to Nature and to natural relations. This return, however, will only become possible when man sees and understands his true "place in nature." He will then, as Fritz Ratzel has excellently remarked,(47) "no longer consider himself an exception to natural laws, but begin to seek for what is lawful in his own actions and thoughts, and endeavour to lead a life according to natural laws." He will come to arrange his life with his fellow-creatures—that is, the family and the state—not according to the laws of distant centuries, but according to the rational principles deduced from knowledge of nature. Politics, morals, and the principles of justice, which are still drawn from all possible sources, will have to be formed in accordance with natural laws only. An existence worthy of man, which has been talked of for thousands of years, will at length become a reality.

The highest function of the human mind is perfect knowledge, fully developed consciousness, and the moral activity arising from it. "Know thyself!" was the cry of the philosophers of antiquity to their fellow-men who were striving to ennoble themselves. "Know thyself!" is the cry of the Theory of Development, not merely to the individual, but to all mankind. And whilst increased knowledge of self becomes, in the case of every individual man, a strong force urging to an increased attention to conduct, mankind as a whole will be led to a higher path of moral perfection by the knowledge of its true origin and its actual position in Nature. The simple religion of Nature, which grows from a true knowledge of Her, and of Her inexhaustible store of revelations, will in future ennoble and perfect the development of mankind far beyond that degree which can possibly be attained under the influence of the multifarious religions of the churches of the various nations,—religions resting on a blind belief in the vague secrets and mythical revelations of a sacerdotal caste. Future centuries will celebrate our age, which was occupied with laying the foundations of the Doctrine of Descent, as the new era in which began a period of human development, rich in blessings,—a period which was characterized by the victory of free inquiry over the despotism of authority, and by the powerful ennobling influence of the Monistic Philosophy.

FOOTNOTES

1 With the exception of a single specimen of the bones of a foot, preserved in the cabinet of Amherst College.—E. R. L.

2 The primary stock of the Coniferæ divided into two branches at an early period, into the Araucariæ on the one hand, and the Taxaceæ, or yew-trees, on the other. The majority of recent Coniferæ are derived from the former. Out of the latter the third class of the Gymnosperms—the Meningos, or Gnetaceæ—were developed. This small but very interesting class contains only three different genera—Gnetum, Welwitschia, and Ephedra; it is, however, of great importance, as it forms the transition group from the Coniferæ to the Angiosperms, and more especially to the Dicotyledons.

3 "Ueber ein Aequivalent der takonischen Schiefer Nordamerikas in Deutschland."

4 The English word "Insects" might with advantage be used in the Linnæan sense for the whole group of Arthropods. In this case the Hexapod Insects might be spoken of as the Flies.—E. R. L.

5 Weisbach: "Novara-Reise," Anthropholog. Theil.

6 Müsset im NaturbetrachtenImmer Eins wie Alles achten.Nichts ist drinnen, Nichts ist drauszen,Denn was innen, das ist auszen.

7 Natur hat weder Kern noch Schale,Alles ist sie mit einem Male.

www.ingramcontent.com/pod-product-compliance
Lightning Source LLC
Chambersburg PA
CBHW051906170526
45168CB00001B/270